CUTTING
Loose

ASHTON APPLEWHITE

CUTTING
LOOSE

WHY WOMEN
WHO END
THEIR MARRIAGES
DO SO WELL

HARPER

NEW YORK · LONDON · TORONTO · SYDNEY

HarperCollins
PUBLISHERS
Since 1817

FIRST HARPER PERENNIAL EDITION PUBLISHED 1998.
REPUBLISHED IN HARPER PAPERBACKS IN 2017.

Designed by Alma Hochhauser Orenstein

The Library of Congress has catalogued the hardcover edition as follows:

Applewhite, Ashton.
    Cutting loose : why women who end their marriages do so well /
Ashton, Applewhite. —1st ed.
        p.   cm.
    Includes bibliographical references and index.
    ISBN 0-06-017455-2
    1. Divorced women—United States—Case studies.   2. Self-realization—United States—Case studies.   I. Title.
HQ834.A72   1997
305.48'9653—dc21   96-51896

ISBN 978-0-06-268070-9 (pbk.)

HB 09.09.2022

*To all the women who so generously trusted me
with their stories*

*and to the memory of my mother
Joyce Zinsser Applewhite
1920–1996*

Contents

Acknowledgments

For their love and support I thank my daughter, Morgan, and son, Luke, who had to live with this project; my father and mother, Ed and June Applewhite, who were *always* interested; Bob Stein, for ideology checks, umbrage, and unwavering confidence; my sister Maria and brothers Jarratt and Anthony, for bearing with me through the divorce and beyond; and Ruth, Bill, Katie, Murphy, and Patty Stein, for encouragement and forgiving me for working through the weekends.

My greatest debt is to Nancy Peske, without whose editorial insight and energy this book would have been longer, later, and nowhere near as good.

For early readings, I thank Andrew Frothingham, Clyde Taylor, Margie Allen, Rachel Patron, and Brian Drolet; for professional guidance generously tendered, I thank Pam Dorman, Stuart Krichevsky, Dawn Drzal, Victoria Segunda, Terry Karten, Kera Bolonik, and Flip Brophy; for expert readings I thank Don Schuck and Martha Beuerle. I also thank Kathleen Drolet for that morale-boosting telephone message; Amity Janow for sending me Congreve; Lynn Decker for the Word tutorials and clippings; Lindy Hess for confidence sight unseen; Lisa Pedicini for the champagne and Post-it; and Charlotte Kreutz for those spring phone calls.

Preface

One of the low points in my life occurred thirteen years after this book was first published, in June 2010. My extended clan had gathered at the White Horse Pub in Burnham Green, a suburb of London, the night before my stepdaughter Katie was to marry a wonderful British man. It was toast time. My twenty-three-year-old son Luke, three years younger than Katie, stood up and said, "My parents divorced when I was seven, and it really messed me up."

Guilt flooded through me once again at having upended my children's lives. I managed a blank look and resisted the urge to crawl under the table as Luke kept talking, describing upheaval and transitions, family trips and sibling rivalries, resentments and rewards. After what felt like an hour, he brought it home: "If that divorce hadn't happened, my life would have been so different. I wouldn't be here now, I wouldn't have Katie in my life, or Alex, or Penny"—Katie's fiancé and their infant daughter—"and I'm so glad and so grateful that this is the way things turned out."

The point is not that our family is lucky, although we certainly are, but that it's impossible to envision how things will turn out—

especially when we're in the thick of it. I could never have foreseen that my life would parallel Laurie's, a woman interviewed for this book who recalled with horror her son, Sean, yelling, "You're ruining my life." When she went on to report that Sean now had a great relationship with his dad, an affectionate stepfather, and two half brothers who worshipped him, I felt a fierce pang of envy.

Bob's and my families are blended in ways that seemed unimaginable when I wrote this book and we were new to each other. It's Bob's ex-wife Aleen, always more family-oriented and fundamentally nicer than I, who calls for family get-togethers whenever she comes through New York. I don't hold a candle to her in the grandchild-diapering-and-feeding department, but no one's asking me to, and being a working *grand*mother is a lot less stressful than being working mom. Our children and grandchildren have more models to learn from, adopt, or reject. What could be better?

It's been almost a quarter century since *Cutting Loose* came out in 1997. What has changed for American women and their families during those decades, and what—for better and worse—has not? Back then the *Yellow Pages* were still around and searching the internet meant using a modem to log on to America Online. Conservative crusader Newt Gringrich was Speaker of the House and Jerry Falwell's Moral Majority was trumpeting the need to return to "traditional" family life. "The current family values mania is part of a last-ditch effort to find security in an era of shrinking personal and economic security," I wrote, "and to restore to men a clear-cut definition of masculinity."

Sound familiar? During his 2016 campaign, Donald Trump named Gingrich a top candidate for the vice presidency. Trump's core constituency, working-class white men, continue to be buffeted by economic and social forces that are reshaping their place in the world. Demographic shifts have turned them into a racial minority. Globalization has further eroded their role as breadwinners. A man's home stopped being his castle decades ago. Now he's lucky if it's a tract house and there are two incomes to cover the bills.

These changes have also rippled through women's lives. In 1950, women represented less than 30 percent of the workforce; by 1980 we made up more than half of it, and by 2014 it was 57 percent. Although this rate is predicted to slow markedly, the Great Recession required many more women to find jobs in order to pay their families' bills. Not only are more women working outside the home, wives in dual-earner couples are contributing a greater share of the couple's total earnings, 40 percent in 2011, up from 38 percent in 2006. And more wives in 2011 were out-earning their husbands.

When both partners "bring home the bacon"—or the free-range chicken—decision-making, child-rearing, and domestic tasks tend to be shared more equally. Men are doing more of the housework than they did when this book came out, and spending more time caring for children, though still only about half as much as moms do. Parental leave continues to be cast as a women's issue instead of a family issue, although this should change as fatherhood becomes more integral to more men's identities and paternal leave as common as maternal leave. In other words, we're on the right track but still a long way from the gender equality in domestic life.

We're still a long way from gender equality outside the home too. A woman's access to safe and legal abortion remains in jeopardy. The glass ceiling remains firmly in place. The wage gap kicks in at a shocking age, thirty-two, and remains stubbornly persistent. In 1992, women working full time, year round were typically paid seventy-two cents for every dollar paid to their male counterparts. Almost a quarter of a century later, in 2015, women still made only eighty cents for every dollar earned by men. The effects mount up over time. The Institute for Women's Policy Research estimate that the wage gap costs a college-educated woman almost $800,000 by the time she turns fifty-nine. The gender pay gap is worse for mothers. Men's wages continue to go up when they become dads; women's wages go down when they become moms. The long-term consequences for women's health, autonomy, and well-being are devastating, and are hugely responsible for the fact that the subtitle of this book continues to surprise.

What about the long-term trends in place at the end of the twentieth century: later marriage, fewer children, more divorce, more remarriage. Are they still trending? Mainly yes, partly no. The average age at which both men and women first marry has continued to inch upward, twenty-seven for women and twenty-nine for men, giving people more to get to know themselves and to figure out what they want in a partner—at least in theory. More people now delay marriage, waiting until their thirties or forties. Waiting to marry until one's early thirties used to raise the chance of divorce; now it lowers it. That's partly because so many people now live together before getting hitched, which means that many relationships end in break-ups rather than actual divorces, or never lead to marriage at all. The stigma of illegitimacy has largely disappeared, mercifully taking the shotgun marriage along with it—never a solid foundation for life partnership.

Americans are having fewer kids, and citing the rising cost of child-rearing as the main reason. The statistic that kicked off this whole project—that two thirds of divorces are initiated by women, which took me completely by surprise —is holding steady. Divorce rates peaked in the 1970s and 1980s and have been declining, with those who married in the 2000s divorcing at even lower rates so far, although the rate of "gray divorce" (over the age of fifty) has doubled since 1990. If current trends continue, nearly two-thirds of marriages will never involve a divorce.

That says nothing, however, about your particular circumstances or reasons for picking up this book, and it's only true of part of the population. While divorce rates have mainly declined for college-educated couples, working class couples are more likely to split up at near-peak levels. "Better-educated Americans have found a new marriage model in which both spouses work and they build a strong economic foundation for their marriage," explains sociologist Andrew Cherlin, author of *Labor's Love Lost: The Rise and Fall of the Working-Class Family in America.* The erosion of the middle class, on the other hand, has left many of the less educated struggling to find work and more vulnerable to splitting up during economic hardship.

Serial transitions in and out of long-term relationships are now typical of family life in the United States, which made stepfamilies more common. Approximately one-third of all weddings in America today create stepfamilies. More than four-in-ten American adults have one step-relative in their family—either a stepparent, a step or half sibling or a stepchild, according to a nationwide Pew Research Center survey.

Aleen had two children when she married Bob, and Katie set me straight very early on when I referred to one of them as her half-brother. "We say 'brother,'" she told me gently but firmly. (Just to keep things confusing, his name is Morgan and so is her sister's—the one she acquired when her dad met me.) When I ran the wedding-toast story past her for corroboration, Katie replied, "The kids didn't like the feelings they had in the process of divorce, that's normal. But our big and weird and inclusive family is one of a kind and I don't think any of us would change it for the world."

Blood exerts a primal pull, but it doesn't guarantee healthier relationships. Changes in family form, advances in reproductive technology, longer lives, and the embrace of "families of choice" have all combined to demote biology as the organizing principle of family life. That's a good thing, because it's more realistic, more forgiving, and more inclusive. My physical therapist, a nice Catholic girl, was flipping out about turning thirty-five, unmarried and childless, until she started working at New York's 92nd Street Y with some of its legendary older women members. "These women were in their third, fourth, and fifth acts," she marveled. "They belonged to all kinds of families. Actually having children, or husbands, didn't have much to do with it." What's a family in the twenty-first century? Any group of two or more people committed to looking after each other for the long haul.

Relationships are likewise assuming more varied forms in America. More of us are cohabiting, raising children as single parents, and living alone than ever before, with almost 30 percent of American households now consisting of just one person.

In 2012 the share of American adults who've never tied the knot hit an historic high, at one-in-five adults ages twenty-five and older. (In 1960, in comparison, the percentage was one-in-ten.) In parallel, less than half of all adults in the United States are currently married. That tipping point occurred in 2011, when the percentage dropped to 48 percent (compared to a whopping 78 percent of American households in 1950). Pushing twenty-five years together, Bob and I are part of that unmarried majority. We got domestically partnered for health insurance reasons, but one trip down the aisle was enough for us.

Nevertheless, most people eventually marry, especially the better-off. Married people continue to enjoy a slew of legal and logistical advantages, from Social Security benefits to inheritance laws, and many people value the institution for religious reasons.

And even as the number of married people declines, the number of those who remarry has been rising for decades. In other words, previously married people are as willing as ever to give matrimony another shot. That includes three out of four divorced people (although, as ever, the probability that the union will end in divorce increases with each trip down the aisle.) What's driving that trend? Divorced rates are high enough that plenty of Americans are candidates for remarriage. Longer lives mean that a growing number of widows and widowers are eligible too.

Perhaps the strongest evidence that marriage remains central to American life has been the movement for marriage equality. On June 28, 2015, the US Supreme Court ruled that no American could be denied the right to marry because of their sexual orientation. Justice Anthony Kennedy explained his vote by declaring marriage to be "essential to our most profound hopes and aspirations." My deepest hopes and aspirations don't have much in common with Justice Kennedy's. Of course gay people should have the right to marry, but marriage should not be the gateway to social and economic privilege. I hope we continue to move towards a society that supports the full range of relationships and family forms—married and unmarried, straight and gay, monogamous and polyamorous, fertile and childfree—and values them equally.

For most of its history, as this book describes and as historian Stephanie Coontz has explained in depth, the institution of marriage mainly benefited husbands and mainly served the patriarchy. That's still the case in the traditional, husband-headed model championed by religious conservatives, and many marriages come apart under the strain of trying to adapt to the more egalitarian gender roles that are clearly the way of the future. Those trends surely underlie a sharp age division in public attitudes towards the value of getting married and breeding, with young adults more likely than their older counterparts to say society is just as well-off if people have other priorities. The future of marriage depends on whether the institution can adapt to changing gender norms and whether both women and men can calibrate their expectations accordingly.

As I wrote in these pages almost twenty-five years ago, "Until men and women are paid the same wages and are jointly responsible for family life, there will be no such thing as equal opportunity. . . . Until men and women come to the table as equals—be it in the kitchen or the boardroom—as true peers, the egalitarian marriage will remain the exception, if not a downright contradiction of terms." It's still damn hard to have an egalitarian marriage because our society still values men and women differently, but the gap is closing—albeit not fast enough.

It is still a sexist, patriarchal world. Structural discrimination against women remains firmly entrenched, affecting identity, health, career, and financial security. Caregiving, largely work done by women, and increasingly by poor women of color, remains underpaid and undervalued. The effects cascade into family life and compound across the life course, disadvantaging women lifelong and leaving many impoverished in old age.

It's up to all of us to work to close that gap: husbands as well as wives, gay partners and straight, coupled and going it alone. When we make the world a better place to be a woman, we also make it a better place to be queer, to have a disability, to be older, to be human. As the poet Audre Lord wrote, "There's no such thing as a single-issue struggle because we don't lead single-issue lives." It'll be a world in which the guidance and companionship you'll find in these pages will be far less necessary—and wouldn't that be great?

Ashton Applewhite
Brooklyn, New York
October 2016

Introduction

*I*n the fall of 1992, after ten years of marriage, I sued for divorce. It was an agonizing decision. My parents were pushing fifty years together. None of my sixteen cousins had divorced; the only family renegade was my notoriously belligerent great-grandmother Frida, who booted hapless William out of her Central Park West apartment in 1917. When I married, on a beautiful May afternoon in Washington, D.C., I was positive it would be forever.

We were very lucky in many ways. Just before getting married I had written a book that became a best-seller, and when it turned into a series my husband quit his job in sales promotion to go into business with me. With the proceeds we bought a nice apartment in Manhattan and for a long time had no money worries. Our daughter was born in 1985 and our son in 1987, and they are delicious children. I had the luxury of spending as much time with them as I wanted to, and for a long time was busy with babies and books.

I was the breadwinner, the mom, and a more productive and ambitious person than my ex-husband. Unfortunately, he was both deeply insecure and extremely competitive. To compensate I tried in any number of ways to "make myself small," as I came to think of it—not because he demanded it but because it conformed to some inarticulate notion of mine of how wives should behave.

It took years to acknowledge the depths of my anger and resentment and depression. Therapy helped. Gradually my eyes opened to this diminished version of myself, tiptoeing around a man who couldn't even afford to like me. Only bit by bit did I realize how much of myself I had voluntarily signed away in the service of some abstract, vaguely romanticized ideal that had nothing to do with my best interests, or even those of the marriage. My sister's husband was dying that winter, and as I wept for him I eventually realized I was mourning the death of my marriage as well. Not long after our tenth anniversary, my husband moved out for a trial separation. When he announced that he was moving back in, I knew that our problems had not been and could not be solved, and said I wanted a divorce.

Early on in the process, a chance comment by my lawyer set me to thinking. "Most of my clients are people like you," he said, "wives who realize they don't have to stay in unhappy marriages." I was intrigued. I had headed into my divorce with the vague but distinct impression that it left women financially devastated and socially shipwrecked. Plentiful articles in the popular press made it clear that the chances of a woman over thirty finding a new mate only slightly beat the odds of getting struck by lightning, while ex-husbands chose successors from a shelf of trophy wives. Well-publicized figures showed a divorced man's income soaring while his ex-wife's took a nosedive, and until I read Susan Faludi's *Backlash*, I believed them.

If this was indeed what lay ahead, how come wives have always sought divorce in greater numbers than their spouses, two to one at the turn of the century, almost three out of four (71 percent) in 1928, and 75 percent in the 1990s? Had my predecessors, like me, believed this grim prognosis but decided to take their chances anyway? Could it be that the consequences weren't actually all that dire? I resolved to find out . . . as soon as the wrenching logistics of my divorce had been worked out. It took a relatively speedy seven months, an eon to me, and cost me $11,000.

Our property was divided fifty-fifty, as are the children's expenses. They spend three nights a week in the apartment my ex-husband moved into two blocks away. Though the children clearly miss the four of us being together, they are thriving. If I'd been willing to haggle longer or go to court, I probably could have gotten a better financial deal, but I'd do it again tomorrow. I know something precious has been lost, but also that much of what I mourn was illusory. I used to torture myself with a mental Polaroid, ripped in half, of the four of us at Yellowstone Park. It took me quite a while to admit that I never could have gotten my ex-husband anywhere near that hypothetical Winnebago, and that the perfect family my children will always yearn for never did exist. They are better off now, and so, by far, am I.

Although my expenses are higher, my standard of living hasn't changed because I began taking professional risks that have paid off. I'm deeply involved with a wonderful man who delights in my competence and understands why remarriage holds little appeal. Best of all, I feel great: in full possession of myself, responsible, sexy, independent, and powerful.

Society doesn't look kindly on this sort of transition. The idea of a woman opting to be husbandless is profoundly threatening. My family was terrifically supportive during my divorce, but not long afterward my fierce and wonderful godmother declared me to be the most selfish person she'd ever known. Why? Because I was ambitious. I had never lost sight of my professional goals, even now, as the unmarried mother of two young children. When I accused my mother of tarring me with the same brush, she protested, "I never called you selfish. I said *self-centered*." Along with annoyance that a man would never have to fend off such an accusation came the exhilarating realization that this pronouncement no longer wounded me. I had indeed learned to look after my own needs. I wasn't self-centered, I was centered, and it had taken a divorce to get me there.

Marriages are mysterious. Many please and fulfill their partners, and many of the women in this book are very happy in second marriages in which they are more equal partners. The crux of the problem is that maintaining such a marriage in a world that values women's experience less than men's is enormously difficult. "Are you sure most people see marriage as a partnership between equals?" a friend asked flatly. His question is a good one. Lots of people—like the man who advertised in the *Parade* Sunday supplement for an "old-fashioned wife," and the woman who responded and in due course promised to love, honor, and obey him—don't aspire to any such thing. I hope they live happily ever after, but in our rapidly changing society it's not going to be easy.

Cutting Loose is a book based on interviews with women who initiated their divorces. The interview criteria were simple: subjects had to be female; they had to be the ones who initiated the divorce; and the papers had to have been signed. Word-of-mouth, notices in publications ranging from the *Pennysaver* to the *Romance Writers Report*, and letters to women's and divorce-related organizations produced an abundance of subjects. I was struck by their resourcefulness, their courage, their willingness to come forward with stories by turns moving, daunting, hilarious, and highly instructive. I cherish their trust and their words. Once, flying back from California with precious tapes stashed in my bag, I had the fleeting sensation of feeling pregnant with their stories.

Some of the women I interviewed have remarried while others have stayed single; some earn six-figure incomes while others scrape by; some are fierce and others gentle. Subjects vary in age, income, and background; in the duration of their marriages; and in how much time has passed since the divorce. A few were married to monsters, but most to decent men who also tried hard to make their marriages work. Most are mothers, but some are childless; a few wished all aspects of their identity to be concealed while others felt comfortable using their real names. It's a self-selecting sample, with no claim to being statistically representative. While each of these women seemed very different from me and from one another, I was constantly startled by echoes of shared experience. Race and class, fault lines along which politics and theory usually divide sharply, did not segregate these women's tales of pain and growth. Each of us had at some point acknowledged how compromised we were in our marriages, and had resolved to do something about it.

Cutting Loose is not a how-to manual; plenty of those already crowd the shelves, and who reads the safety instructions anyway? The focus is not on individual narratives; every marriage is two stories. (To trace a single narrative, consult the "Index of Women Interviewed" at the back of the book.) Rather, the book is a collective portrait, organized around the themes that emerged as these women talked about their experiences, reflected on what they've lost and gained, and described their lives now. In their own words they tell why they decided to make the change, what getting divorced is really like, what moves they're proudest of or most regret, how their feelings have changed over time, and what advice they'd give now to women contemplating divorce. After all, where would women be without other women to talk to? This is a book about women taking charge of their lives and emerging stronger, clearer, and infinitely happier. It's the book I wish I'd had before I myself cut loose.

In my very first interview, Karen, a forty-nine-year-old sculptor who's been married and divorced twice, commented:

> Since I've gotten divorced, or have been in this process for the last four years, a lot of my friends have been in the process. Both younger and older friends, and people that I work with and know. They seem to fall into two categories: either people who have stayed in marriages and made tremendous compromises in what they wanted their lives to be like, or people who have ended their marriages and also gone through tremendous compromises in terms of what they wanted their lives to be like, in terms of the economics of it and that sort of thing. But I think that the people who decided to take the responsibility for their lives, outside of marriage, are happier, less compromised, less bitter.

She'd come to the same conclusion as the one I soon arrived at: women who end their marriages are far better off afterward.

This despite the economics of divorce, which are indeed punishing, especially for mothers of young children. But even Lenore Weitzman, dubbed the "godmother of the backlash against divorce" by *Time* magazine, had trouble finding divorced women who regretted the change. In *The Divorce Revolution*, she wrote, "Even the longer-married older housewives who suffer the greatest financial hardships after divorce (and who feel most economically deprived, most angry, and most 'cheated' by the divorce settlement) say they are 'personally' better off than they were during marriage. . . . They also report improved self-esteem, more pride in their appearance and greater competence in all aspects of their lives." Like my subjects, though not by choice, these women came to realize that traditional marriage serves the husband, the wife serves the marriage—and that independence beats servitude.

These women's attitudes are all the more noteworthy given the tenor of the times: they sure aren't getting any support from Congress or the media. Spokesman of the New Right Newt Gingrich routinely touts a return to "traditional family values." The beleaguered American family could use some real help, though nothing Gingrich would approve of: better day care, a corporate culture amenable to flextime and job sharing, and equal pay for men and women.

Jerry Falwell, head of the Moral Majority, declared, "Good husbands who are godly men are good leaders. Their wives and children want to follow them and be under their direction." The women in this book know otherwise: though it may not be their first choice, women can do just fine without husbands. Often better, when the alternative is a relationship in which mutual respect or goodwill are bitterly absent. The word "divorcée" still conjures up an image of a sad and hungry homewrecker on a barstool, a woman who failed to put her husband and family first and got what she deserved. The characterization of divorced people as irresponsible and immature—what Dr. Constance Ahrons calls "divorcism"—persists. The New Right doesn't want to let her go unpunished, because it is far more threatening to America's "family values" when divorced women *don't* fall apart, when, with or without a new mate, they carve out lives of unparalleled productivity and richness, on their own terms.

In *Composing a Life*, a celebration of the improvisatory nature of women's lives, writer Mary Catherine Bateson says, "I have come to believe that when the end comes it pays to cut your losses, for there is almost always more ahead than we can guess." She's right. *Cutting Loose* shows women what it takes to make the move—the same strength it takes to keep unhappy marriages going—and describes a complex and exhilarating future. Over dinner not long ago, I described the book to a British journalist seated next to me. "Isn't your book going to be awfully, you know, depressing?" he asked solicitously.

"Oh no," I replied with a grin, "no, not at all."

CUTTING
Loose

CHAPTER 1

Shattering the Illusion

> . . . if I continue to endure you a little longer,
> I may by degrees *dwindle into a wife* [italics added].
>
> –MILLAMANT, IN *THE WAY OF THE WORLD* BY WILLIAM CONGREVE

What happens when women turn into wives? Too often they dwindle, which is why Congreve's phrase, written in 1700, reverberates so clearly today. A snippet from the opposite end of the literary spectrum, the gossip column of the *New York Post*, observed that the two wives of New York financier Sid Bass had apparently traded identities. "Mercedes, the current Mrs. Sid Bass, has metamorphosed from uninhibited and fun-loving to tense and chilly. Whereas Anne, the former Mrs. Sid Bass and the original ice princess, is now the outgoing one, the one you want to sit next to." The same topic came up when Tory, a human resources coordinator, was talking about wives with her best friend one day, "about how there's something that bleeps off their radar screen when they get married. They started out being bright people, and some element of their view of the world kind of drops off."

Dwindling into Wives

Marriage reduces many women, who willingly, often unthinkingly, embrace a peculiarly circumscribed identity and set of priorities when they give up being single. Looking back, they are often frustrated and puzzled by their own collaboration in the process. Susannah, at forty-nine a respected film archivist and the kind of person you can't imagine taking no for an answer, married a much older artist who shaped much of her taste and opinions. She describes her married self as "kind of paralyzed. I couldn't make decisions." Jodie, a widely respected software industry spokesperson, admits, "I had so surrendered to him and to his way that I didn't know who I was anymore."

For my part, I worried constantly about incurring my husband's displeasure, an unnamed fear way out of proportion to any possible consequences. No brute, he was handsome and funny, and we had a lot of fun. We were true to each other. We seldom fought. He said no to many of the things I wanted to do, which angered me, but instead of acknowledging it I colluded vigorously, even frantically, in the illusion that I was getting as much out of the marriage as he was. It held up for a long time, especially during the years when I was busy with babies, just as it did for Karen, a sculptor, who says that her divorce was made far harder because "it took me so long to realize my own investment in this fantasy."

Now I wonder how on earth I reconciled my strong, articulate self with that anxious, muted creature. What I didn't understand was how the illusion that he and I were equal is built into our culture. The truth is that the only time when a woman's social worth equals a man's is during courtship, when the man must work to win her. Small wonder that little girls dream of being brides, the pinnacle of desirability, the ultimate female leading role. But Bridal Barbie offers nothing but a short-term sugar rush. Once married, no longer an object of competition among the men, a woman finds that her value plummets. "It's like a new car," one woman commented. "The minute you take it out of the showroom, it loses 25 percent of its value."

Consciously or not, many new husbands feel justified in treating their new wives less well, or at least reminding them of their diminished status. Many brides noticed that romantic behavior abruptly stopped after the wedding. One husband declared dancing no longer necessary; another pulled the plug on kissing and oral sex. Anneke, a romance writer from Tucson, doesn't think the change in men's attitudes is conscious, but rather that "for the most part once they're settled and secure in a situation, they just don't try hard." When she and George were dating he indulged her fondness for picnics and movies, but six months or so after the wedding, the balance shifted, permanently. "I really think that as soon as he was sure of me, he just put down the mask," declares Anneke. "I've seen it in so many marriages, where the first year or two he does everything, takes out the garbage, God knows what all, and then after a while he just comes home and plops in front of the TV set and says, 'Bring me a beer.'"

Promising to have and to hold no matter what, a bride willingly hands over a chunk of her adult identity in return for what is supposed to be an enduring romantic partnership. It's a time-honored exchange: social and legal autonomy for the promise of intimacy, the bulwark against solitude, and the thrilling challenge of making love last. Security, status, and silverware will plaster over the rift between what's fair and what she ends up with, and compensate if romance ebbs away.

Although one out of every two marriages ends in divorce, women still aspire to wed, and despite the feminist gains of the 1970s, they still line up to buy the bill of goods that comes with "I do." They buy into the idea that marriage and motherhood are the main routes to fulfillment, that having a career and a community of friends is not enough, that a woman alone is somehow deficient. (The bachelor, on the other hand, is an object of envy—go figure.) They discount as absurd the possibility that a woman might actually *choose* to be single. They accept the definition of "women's work" as tasks that women are naturally suited for, rather than as what it really is: drudgery that men don't want to get stuck with. They believe that women are selfish if they go after what gives them value: the same control over their physical and intellectual lives that men have. Women marry because wives are supposed to have cornered the market on womanly happiness. In fact, wives constitute the most depressed segment of the population (and women who describe themselves as happily married suffer nearly four times as much severe depression as happily married men).

Some people believe that the cure for the troubles of the American family is to go back to the good old days of *Leave It to Beaver*. In the traditional arrangement embodied by June and Ward, the wife tends to heirs and housework, leaving her husband free to pursue his own career and hobbies, or to be a domestic hero and change some diapers. She lets him "wear the pants," and never checks the closet for a second pair.

One of the problems with this traditional arrangement is economic. In the downsized nineties few couples, let alone families, can get by on one paycheck. If the husband makes enough to support the family single-handedly, the economics of the traditional marriage work: he has his job and she has hers. In theory, the contributions are equally valued and the arrangement satisfies both partners. In reality, paid labor gets more respect than the volunteer variety, and living through children or husbands erodes the sense of self. Amnesia abounds; Betty Friedan figured all this out forty years ago in *The Feminine Mystique,* her study of the suburbanization and isolation of the American wife. Returning to the *Father Knows Best* era won't solve anything.

The illusion is that women need marriage, when in fact it's the other way around. The women in this book came to realize that they had given up something ineffably precious and received little of equal value in return. Strong women, all had nevertheless abdicated a certain central responsibility for themselves, and this bad bargain ate away at them.

"I was supposed to be there to bring him his slippers and pipe."

Married in 1955, Olivia very much wanted to be a loyal and loving wife. When times changed and her husband didn't, she learned how hard it was to break free of that traditional and ultimately demeaning role. Now almost seventy, she is a 1960s spirit—peace marcher, inner city volunteer, women's rights advocate—whose wedding turned her Ivy League education into an irrelevance and her ambition into a handicap. Olivia shakes her cap of thick gray hair and puts her sandaled feet up on a footstool as she describes her mother making clear to her daughters what their destiny was to be. When Olivia's sister announced that she wanted to go to medical school, her mother responded, "Oh no, dear, you want to marry a doctor." Olivia distinctly remembers her state of mind on the eve of her wedding to Duncan, a solid, conservative businessman. "I sort of had the world by the tail: I had a good job, I had a great man to marry, everyone was healthy, and I was *really* feeling great, like I was standing on tippytoes. And I remember saying to myself, 'Hang on to this, if you possibly can, in a marriage.'"

The feeling was fleeting indeed. After Olivia and Duncan's small family wedding, friends came up to their room in Chicago's Palmer House to celebrate. When one of them started horsing around with a bowl of popcorn to loosen things up, Duncan picked up a pear from a bowl of fruit and smashed it on the offending guest's head. "Whereupon, of course, everyone got up and said, 'I think it's time for us to go,'" recalls Olivia, "and there I was feeling properly subdued, properly scared." That's when the feeling of having the world by the tail evaporated, "right then and there," she admits, a rueful grin spreading across her calm, uncomplicated face.

Olivia stopped working when the first of her five children was born. Though the marriage lasted for thirty-five years, Olivia maintains that she was "never really made to be a traditional wife in the way our culture demands it. Never. I was too independent. I would get passionately interested in things other than my husband and that was a problem." Her volunteer activities were fine with Duncan as long as dinner was on the table promptly, as long as the phone didn't ring, as long as Olivia was home all evening. The inevitable interruptions would make him so furious that Olivia would eventually give in and quit the offending committee. But once the kids were in college, she opened an art-postcard store with a partner, and when it flourished, she dug in her heels. "I was working, and that bothered him. His business was in the red, ours was in the black, and that *bothered* him," she says. "But I was *not* going to quit *this* because of Duncan. It was too important to me."

The moment when Olivia actually decided to leave the marriage is crystal clear in her memory. After a meeting about a line of greeting cards, she invited the sales rep to stay for a glass of wine. They were on the terrace when Duncan brought his marketing director home to have a beer and run some numbers. The minute both guests left, Duncan turned on his wife in a rage.

He said, "I come home from work and I find you on the terrace, having a drink with some bimbo?!?" I said, "Well, it's exactly the same thing for you, what's the difference?" He couldn't explain it; he just didn't expect me to be doing that. I was supposed to be there to bring him his slippers and pipe. At that point I realized, clear as day, that there it was: there was never going to be equality between us. He couldn't even conceive of it; in fact the idea of it made him *furious*. So I decided at that moment that I'd had enough. It was a little thing, but there's always a breaking point.

"I had to make myself small so that he would feel big."

Married twenty-five years later than Olivia, a child of the seventies, I too struggled to maintain the traditional balance of power in my marriage. Since it meant balancing an imbalance, it was an impossible task. To appear submissive, a role I chose and embraced and had no talent for, I fell into the pattern of asking my husband's permission: to rent a car, to have the So-and-so's over, to leave the kids with him and go out with a friend. To maintain the authority we both felt was appropriate but sensed was precarious, he'd say no or "I'll have to think about it." I would calculate how and when to present decisions in the hopes of getting a yes; he would see right through me and call me manipulative. Eventually I came to see it as making myself small so that he in turn would feel big. If he felt in charge, I reasoned, he would be more generous in spirit. It didn't work, because underneath, though we never talked about it, we both knew better.

In a more egalitarian relationship, or society, my choice—to diminish myself so as to maintain the illusion of balance—wouldn't have been necessary. My behavior was rooted in a cultural assumption that empowering women disempowers men, and reinforced by the fact that, like many men, my husband felt threatened when decision-making was shared. Many husbands feel that someone has to be on top, and it had better be them.

I buried my resentment for years. I did muster my courage and say to a counselor we saw four years into the marriage, "I think my husband feels that my accomplishments diminish him." Perhaps because he was also my husband's therapist, the counselor never pursued this idea, and my husband and I floundered on. A few years later my husband agreed to my suggestion that we take our then preschool-age children and live in Europe for six months or so—on condition that we go someplace where I didn't speak the language. That would make him too reliant on me, he explained. Since that ruled out France and Italy, I gave up the idea, without even a squeak of protest, not to mention a frank discussion about what was underlying his proscription.

Arlie Hochschild, author of *The Second Shift* and a professor of sociology at the University of California, Berkeley, calls behavior like mine "balancing." In her research she observed that if men lose power over women in one way, they make up for it somewhere else, whether by avoiding the housework (the "second shift" to which her title refers), controlling the purse strings, or withdrawing emotionally. "Women who 'balanced' felt 'too powerful,'" writes Hochschild. "Seeing when their husbands got 'touchy,' sensing the fragility of their husbands' 'male ego,' not wanting them to get discouraged or depressed, such women restored their men's lost power by waiting on them at home." In fact, of course, this behavior reinforces an *im*balance between husband and wife; both partners feel "normal" when she's doing more.

Women with more forceful personalities or clearer agendas than their mates often butt up against domestic scenarios that cannot contain their appetite or ambition. The very attributes they hunger for in their spouses are for them sources of conflict and confusion, and "balancing" is their way of reconciling the disparity. I doubt that my husband consciously expected me to be deferential; the unnerving part is that I simply assumed I should. It was, ironically, a source of comfort in the face of larger uncertainties. This, after all, was how wives behaved. Being a wife was important to me, and mine, after all, was a "modern marriage"—or was it?

"There was a lot of lip service."

Nancy's was a modern marriage too, even though she was raised in Tennessee in a very traditional family, "waiting to find a husband who was going to define me," she says in a voice full of irony. Instead she left the local boys behind, got a degree from the Rhode Island School of Design, swapped the big hair for a spiky 'do and an all-black wardrobe, and moved to New York. She was too busy establishing her career as a tableware designer to contemplate marriage until her early thirties, when she met Eric, an advertising account executive, at a friend's engagement party. They lived together for three years and were married for six. It never occurred to Nancy that she couldn't "have it all": marriage, career, and kids.

The Modern Wife has two jobs, a paying position and a volunteer one. In theory, the husband does half the domestic tasks. In reality, this is almost never the case, even when the wife makes more money. Men do more around the house than they used to, but it's seldom without a struggle, and it only averages 20 percent of daily chores like laundry, cooking, and cleaning. And though the amount of time fathers spent with their children increased from the mid-seventies to the early eighties, it averages out to around one-third of the hours moms put in. Many husbands talk a good game, but many have a hard time being genuinely supportive. A few, like Nancy's, are downright subversive.

Eric was always saying how important Nancy's career was, and was very excited for her when *Metropolitan Home* called about running a photo spread of her new line of flatware. Nancy put together a packet of slides and samples, which Eric offered to mail; two weeks later she found the package on his desk. "I was so devastated that he would sabotage me in that way . . . I knew at that point that I could never trust this person again. And that was very early in the marriage," she says sadly. That's when it dawned on her that she was probably smarter, more talented, and certainly more self-sufficient than her mate. She recalls it as "just a real sobering kind of moment."

> It's like I thought that marriage was a person taking care of you, and I could give everything up and just relax. I look back on it now and I realize very quickly that that was not what this person was capable of doing, and it really wasn't what I wanted to do. So when I started kind of like exerting myself again, he kind of kept pushing me down.

Although Nancy worked full-time, her salary was less than half of Eric's. He was always crying poor and accusing her of not carrying her share of the load. They agreed that their daughter shouldn't be raised by baby-sitters, but when Lauren was born it turned out that Eric expected Nancy to keep working. His lack of support came as a nasty shock. Nancy experienced what she calls "a lot of lip service. He would say, 'You do what's good for you, do your artwork, that's fine with me.' But then when it came right down to it, he wanted me to raise Lauren, clean the house, do the laundry, have a job. He wanted me to do everything. And help him in his profession.'"

Eric wanted it both ways: the income and prestige of a working wife who somehow also managed to be mommy and homemaker. It required Nancy to be both heroic and invisible; she herself didn't matter much. She shakes her head as she acknowledges her part in the process. "I kept kind of putting myself down—it *killed* my self-esteem. It was a very subconscious thing. You kind of act the way you think you're supposed to act."

Women Blame Themselves

"I can't believe I let this guy do this to me, that I just kept taking that."

Because she couldn't manage the unmanageable, Nancy felt inadequate, guilty, and responsible. Therapist and recovery-movement spokesperson Anne Wilson Schaef says, "I have yet to meet a woman who has not acknowledged her own particular fear—of being sick, bad, crazy, stupid, etc.—and embraced it. . . . Frequently women deny the way we see the world or the values we have in order to gain male validation and approval. We resist seeing that what is happening to us is happening because we are female. Instead, we blame our unhappiness and lack of fulfillment on some 'flaw' in our character." That perfectly describes Nancy's behavior, and she has plenty of company. In our culture, somehow, it's always the woman's fault. As writer Kathleen Hall Jamieson observes, "our language is laden with unwed mothers but not unwed fathers, battered women but not battering husbands."

When marital problems come to light, both spouses tend to see them as *her* problems. Eric, like many husbands, refused to go to counseling. "He thinks he's absolutely the smartest thing to ever hit dirt, and that all therapists are stupid," says Nancy by way of an explanation. Women have a culture of self-improvement and men do not. Women take aerobics, seek counseling, go on diets, and plow through countless books that proclaim that if only they would change themselves—access that inner child, wrap themselves in a better brand of Saran Wrap, learn how to express themselves—it could all work out.

Because the woman is almost always the emotional custodian of the relationship, it falls to her to "manage" the inequality—at tremendous emotional cost. Nancy felt as though Eric was always holding a mirror up to her, and that the reflection was "not of a very terrific person." If intimacy is absent the wife figures it's her fault, and most husbands, understandably, aren't inclined to argue the point. If intimacy is conditional on subservience, she may grow resentful and confused, but she hangs on. A wise (and never-married) friend calls this paradox, in which the unempowered person insists that the circumstances are of her own making, the Arrogant Doormat Syndrome.

When Lauren was a toddler, Eric began drinking. He blamed Nancy, saying, "You're making me this way, you're driving me to drink." The moment of truth arrived one night when Nancy feared for the child's safety. "I said, 'Never again. You're out of here.'" The hardest part, especially for women like Nancy who consider themselves competent and assertive, is to acknowledge what's happened and why. "I think that's what made me so mad at myself," admits Nancy. "It's like I can't believe I let this guy do this to me, that I just kept taking that."

"There has to be a reason, another woman, he has to be beating you . . . "

Like both Nancy and Olivia, Megan was raised in a conservative home. One of five daughters, she looks the part of the nice Catholic girl in her pearls, pinstripes, and shining brunette pageboy, but it is her struggle against unthinking conformity that gives her story such poignancy. Right after college, Megan married her high school sweetheart, and promptly turned over the financial decision-making to Anthony. After all, he was an accountant, and his grades were better than hers, so she figured he was better equipped to handle the money. The arrangement suited her fine, except when they had to trade in the big car or move to a smaller apartment because they were living beyond their means.

Observing similar patterns of deferral in other marriages, Megan notes that "even though we're in the nineties and both people work, women are still expected to pick up the dry cleaning and go to the grocery store, to do more than the man does." She attributes this to "growing up in traditional families, where you're not pushed just to be who you are." Loving but inflexible and self-centered, Anthony was more interested in Megan as a cherished possession than as an individual with priorities of her own. He wanted to do everything together, and the tighter his grip on Megan, the harder it was for her to breathe. Megan remembers a skiing trip when he didn't even want her to ski down a different, less difficult trail. "I tell you, I can't go down a black diamond," she remembers pleading. He insisted: "Just do it, just do it."

As a newlywed Megan had grudgingly moved to downtown Boston to accommodate her husband's career. She ended up getting a challenging job doing sales promotion for a growing cable network, and resenting the way her life rotated around the times when Anthony was available to do things that he felt like doing as a couple. Then, two and a half years into the marriage, Anthony decided to go to graduate school. Megan came home one day to find him filling out applications for schools all over the country and simply assuming that his wife would pull up stakes to accompany him. "I just got a promotion! How do you think this is going to affect my career?" she protested. "What career?" was Anthony's response. "He said, 'Meg, all you do is sell little black boxes. What do you think they're going to do, make you a director?' I'll never forget it," says Megan, face pale. "I thought to myself, 'He has no idea who I am, and no respect for me to say that. I have to get out of here.'"

From impulse to action was a hard road. Dutiful, respectful, and good, Megan was mired in the expectations of her family and community. "For years I was terrified to think that I might have a scarlet D on my chest. Oh my God, to say I'm divorced, even get it out of my mouth, was so difficult," she recalls, choking up at the memory. Because she and Anthony were part of the golden couple of their extended family—"the Chuck and Di of our neighborhood," she notes ironically—the prospect of telling her family, not to mention the raft of elderly Italian in-laws, was terrifying. "I'd say to myself, 'I've got to make it work for everybody else, everyone's counting on us to be perfect.'"

"I think still in this society, when people get divorced it's almost not acceptable to say, 'It just didn't work out,'" Megan says. "There has to be a reason, another woman, he has to be beating you, it can't just be that you grew apart." It implies something casual about the decision even when nothing could be further from the truth. She blamed herself for failing to love Anthony, the man everyone figured was "the perfect husband." Megan's moment of truth occurred as she was being wheeled into surgery to remove a huge ovarian cyst.

> I was out in the hallway, all alone, and I had this epiphany that this is how it is in life, that at the beginning and the end I am all alone. And that when I get out of the surgery and I recover in seven weeks, I'm going to make myself happy, because nothing is as horrible or terrifying as going through life this way. I don't care what people say because I'm Catholic. I don't think God wants me to live a lie.

When she got home, Megan told Anthony she wouldn't follow him to graduate school. It was the first time she'd simply said no. In response to his protests that he was only thinking of what was best for them, she said fiercely, "That's bullshit. You're not doing this for us, you're doing this for *you*." The day Anthony left for school Megan found herself weeping as she put away the laundry, out of relief that he was gone, and out of grief because she knew it was the beginning of the end.

It still took almost two years to make it legal, but Megan made the first move toward separation when she insisted they see a marriage counselor over Anthony's Christmas break. He pleaded, "Please don't leave me. I know I haven't treated you fairly, Megan, but *please* don't leave me." The next morning he hugged her and said, "Can this be the first day of our new marriage?" "I said, 'No, it can't,'" Megan answered, face crumpling at the memory. "That was really difficult to say. I said, 'The problems we have can't be erased because we wake up in the morning and it's a new day.'"

Megan had been caught up in the struggle described by Dalma Heyn in *The Erotic Silence of the American Wife*: "If she can annihilate herself altogether and still manage to seem contented, then she has achieved the additionally heroic feat of holding onto her femininity—that elusive quality women are always in danger of losing whenever their selves threaten to burst through all the constraints." It's easier than looking hard for the truth, and it's seductive, because doing for others wins a woman a lot more brownie points—and also requires far less conscious thought—than doing for herself. Stifled for years, Megan had broken the habit of deferral, but had no idea what should take its place, or what kind of person would be living this new life. "It's scary," she acknowledges, "but a scarier thought is, 'What would Megan be like, four years later, if I had not gotten divorced? What quarter of a person would I be?'"

Women Give In

"He didn't really want to take care of me, he wanted to control me."

The flip side of deferral is control, which for one husband after another is an overwhelming obsession. Megan describes Anthony as "a serious control freak." The third marriage counselor my husband and I consulted commented, "He has quite an issue with control." One of the things that attracted Jodie, the software analyst, to her husband was that she thought Paul wanted to take care of her, "but really," she acknowledges sorrowfully, "he wanted to control me." In a post-trial deposition, O.J. Simpson described himself as "a controlling person. I like things the way I like things, and people who come into my life tend to conform to the way I do things." Asked whether his former wife Nicole conformed, Simpson replied, "Yes. Not the last year we were together, but certainly before, she did, yes."

What is it with men and control? America venerates the take-charge type, the self-made man, the entrepreneur, the "master of the universe," in Tom Wolfe's phrase. The more insecure the man, the deeper his need to control, because the alternative—that he may not have all the answers—is so threatening. Unlike a woman, he can't afford to admit to ignorance or helplessness without compromising his masculinity.

If one partner is in control all the time, the other is stuck with an unappetizing menu of behaviors with an uncomfortably feminine cast: bickering, whining, hostile acquiescence. But the desire to be in control isn't a boy thing, it's a human thing. Everyone wants to be in charge: the baby howling from its crib, the teenager with purple hair, the wife who wants to speak and act as freely as the men around her. Deferral isn't natural; women learn it to accommodate men.

"I was the crazy one. 'Reason' would prevail."

Some men are passive-aggressive in asserting control, but Steve, Yolanda's husband of twenty-two years, took conscious pleasure in maintaining the upper hand. She met the handsome, smooth-talking young man on the beach at Cape May, and married him four months later. Steve, says Yolanda, "could look at this beige wall, and if it was convenient for his purposes he could say, 'This wall is orange.' After he'd talked to me for fifteen or twenty minutes, I'd say, 'Oh yeah, it's gotta be orange. I don't know what's the matter with my eyes.'" Raised by alcoholic parents, Yolanda had a habit of deferral that was deeply ingrained. "I was the perfect subject," she admits. "I think that's really important to acknowledge." Asked if she sees a difference between her married self and who she is now, Yolanda replies, "I had no married self." It's hard to believe this of the feisty, animated woman in a tailored pantsuit, smoking Marlboros and looking far younger than her fifty years.

One dynamic that developed early in the marriage to justify Steve's authoritarianism was a tacit agreement that Yolanda was the crazy one. Calling their wives "crazy"—"irrational" is the upscale equivalent—is an important way in which men discount women's opinions and maintain control. "'Reason' would prevail. He won," says Yolanda. "He always won." Steve handled the finances, and Yolanda came to the same conclusion as several other wives in this survey: "there was enough money to do the things he wanted, but not enough to do the things I wanted." She persisted, questioning her husband about their finances, but always got such complicated, drawn-out answers that she would eventually shrug and give up. The constant message that "I simply didn't seem to be able really—did I dear?—to understand basic finances" didn't do much for Yolanda's self-esteem, "and," she says with a laugh, "it's still true: his version I don't understand." (One of the things covered by "his usual wonderfully complex, arcane, mysterious, and mystical explanations" of their finances was Steve's compulsive gambling; a six-figure debt, half of which she had to assume, came to light when they divorced.)

They lived comfortably in the New Jersey suburbs. A civil engineer, Steve earned a good salary and, says Yolanda, was very supportive of her work—as long as her salary was less than his. The public relations director for an educational publisher, she loved her job and with the help of a dependable housekeeper immersed herself in it full-time once her son and daughter were in school.

When their younger child was fourteen, Yolanda's husband began campaigning for a third child. He won, thanks to a tactic she considers downright immoral: involving the two older kids, who began telling her how much they too wanted a baby. When Will was born, Steve couldn't relate to the baby, "partly because he was a difficult child, and partly because his birth had already achieved its objective, which was to keep me around," maintains Yolanda. Steve was impatient and callous, unable to accept that there was something really wrong with *his* son, who turned out to be seriously learning disabled. The alliances shifted, so that it was three adolescents against a mother and child. Yolanda felt isolated and defensive, but her protective instincts enabled her to do for her child what she had been unable to do for herself. "When I realized that his own crazy needs wouldn't let him see clearly what was going on with his own son or be part of the solution, when I finally saw, in my role as a tigress, how damaging this was, that's when I left."

Steve was simply furious. Yolanda's older children, at the time aged twenty and twenty-two to Will's seven, thought she was out of her mind. Now, five years later, "they've both independently said that they completely get it, they don't know how I lasted so long. It's come all the way along. It keeps me happy, you know," she adds with a big smile.

Now, says Yolanda matter-of-factly, "I do my finances like a straight, sane person. At this point I still earn maybe a little more than half of what he earns and I have more savings than he does." When her daughter got married last May, Yolanda was in a better position to help than her ex-husband was, a justifiable source of pride. "My husband was very happy that I was afraid of everything," observes Yolanda. "He enjoyed being the expert, so I was the idiot, because our relationship required those extremes." No longer is she hostage to a situation that depended on her being made to feel inept. "I understand the different ways the world works," she says now. "I have worked out my own life very nicely."

"Submit yourselves unto your own husbands, as unto the Lord."

Some couples solved the issue of control by turning to the Bible, where the roles of husband and wife are clearly delineated. For Hannah the well-defined rules must have been a welcome contrast to her chaotic childhood. Her mother divorced her truck-driver husband when pregnant with Hannah, and had gone through four more marriages by the time her daughter turned eleven. "I think I went into a marriage, and into relationships in general, without a clue as to what they were supposed to look like or what my rights were," says Hannah, understandably. Wisps of dirty-blond hair escape from under the baseball cap that shields blue eyes inherited from Norwegian grandparents; a slender brown hand guides the battered Econoline along the rutted country road.

Barely managing her own life, Hannah's mother had no idea how to handle her headstrong daughter, who was seventeen and living with her grandparents when she met Ed, a student and musician. Hannah dropped out of college not long afterward and married four years later. She was married for ten years and has four children, the first of whom arrived nine months into the marriage.

Hannah happily saw herself as a stay-at-home wife, "a homemaker and a mom. I liked the idea." What she was unprepared for was the way in which she and her needs ceased to exist in the marriage. "I realized how more and more there was no space for me," she observes flatly. "Particularly after having children I didn't have any support from him." Hannah was unhappy about the family's exposure to drugs—Ed played in a band and smoked pot a fair amount—and Ed agreed to change their lifestyle. The couple moved to an isolated farm in the middle of Wisconsin's dairy country and made their living selling organic produce. While Ed attended classes and played music, Hannah tended the farm and home-schooled the children. Ed doled out the bare minimum for the purchases he deemed necessary, which included almost nothing for Hannah herself. Any new clothes came from her sister, since Ed would say they couldn't afford to buy new.

The couple also joined a fundamentalist Christian church. "I don't regret it. I learned a lot," says Hannah, who remains a devout Christian, "but Ed got very much into the dogma of the man being the head of the household. That meant basically that he was God in the house. If he said the kids shouldn't have juice right now then the kids couldn't have juice right now, you know?" Ed had his Scriptures straight: the Old Testament emphasizes the role of the husband and father as lord of the household, and in the New Testament the Apostle Paul instructs wives to "submit yourselves unto your own husbands, as unto the Lord. For the husband is the head of the wife, even as Christ is the head of the church." (Ephesians 5:22–23)

Hannah didn't like his attitude and they argued a lot, but she went along. Geographically and emotionally isolated, she had only her four young children and her faith to sustain her. "There was no sense of having any kind of power over anything, no say, no voice, you know? I didn't feel affirmed at all, in anything." Her clear blue eyes close momentarily at the recollection. As the years went by, Hannah felt more and more invisible. The end came after she went to a club one night to see Ed and his band perform. "I took in the whole scene and I saw this man that was my husband up on the stage with these women who were hardly even dressed, and I thought, 'My God, this is his life away from home—and I don't want anything to do with it,'" Hannah recalls. "I told my sister that later that night, 'You know, I'm done. I don't want to struggle in this anymore.'" Anxious though she was about the censure of the church, it was her faith that gave her the strength to break away from her husband's miserliness and hypocrisy.

Women Grow Silent

"On land it's much preferred/for ladies not to say a word."

When self-interest slips away, and self-esteem along with it, women grow silent. Saying what's on your mind is an important element in creating and maintaining identity, as Dana Crowley Jack notes in *Silencing the Self*. Many a wife, afraid (with reason) that speaking out could lead to catastrophe, learns to keep quiet. In Jack's words, she puts her "feeling self on ice, to be preserved indefinitely, while the duty-ridden 'good self' controls her interactions with her husband." Is the wife's point legitimate? Will it make her husband mad? What is she risking?

Who speaks and who listens reveal much about who matters. After all, language is power, and this power is more conventionally wielded by men. For many women, marriage involves the same bargain that Ariel makes in *The Little Mermaid*. Following the warning of the nasty sea witch, who sings, "On land it's much preferred/for ladies not to say a word," Ariel does indeed trade in her vocal cords for a shot at Mr. Right.

Real debate requires an even exchange of ideas and opinions. When her input tends to be taken personally or seen as criticism, a wife learns to suppress it rather than risk hostility. Patterns of silence are often established early on, reinforced by the husband's anger or emotional withdrawal. Some wives are afraid of their husbands' tempers, but even a woman with no such fear is frequently stymied by her mate's inability or unwillingness to communicate. As linguist Deborah Tannen has theorized, men tend to hear women's complaints as problems requiring them to jump in with one-time solutions rather than as invitations to conversation and closeness, and it doesn't take long for mutually frustrating patterns to establish themselves.

Unaware of their wives' long-term unhappiness, many husbands are stunned and indignant when their wives announce that they really do want out; often men feel that they have been kept in the dark deliberately. One woman recalls her husband saying, "'How come you didn't tell me?' I said, 'I did tell you, I just didn't scream it at you.'" When women do speak, men don't always listen, and when husbands do talk, it can be at their wives, not to them.

"I always had to get him to think it was his idea."

That squelched feeling is familiar to Wendy, for twenty-six-years the silent wife of a prominent Austin cardiologist. She had to phrase things very carefully. "I could not say, 'I think,' 'I want,' 'I believe,'" she recalls. "It always had to be couched in terms like, 'What do you think about a vacation this summer?' 'Why don't you take a look at such-and-such?' I always had to get him to think it was his idea." Now fifty-seven, Wendy married "a hundred years ago," right after her freshman year at the University of Texas in Austin, where she is now a food columnist and cookbook author.

Looking at Wendy dressed in a tailored pink wool suit, gray-blond hair caught at the nape of her neck in a velvet bow, one can easily visualize the pretty sorority sister who married so young. Wendy's daughter got married five years after her mother's divorce, and said bluntly, "'Mom, I am not going to be a doormat wife.' I laughed," says Wendy. "I knew exactly what she was talking about, because I tiptoed around him, and the kids tiptoed around him too. He had the upper hand. I don't know why, although I think there's an authoritarianism that comes out of being a doctor. He just simply had to be in charge."

Speech is assertive. Women feel comfortable talking about their feelings, but a strong opinion infringes on the traditionally male domain of rational thought. It's one thing to say, "I'd really like to start saving for a new sofa," another to assert, "We need a new sofa. Let's start a fund this week." When indirect statements have no effect, women often put off tackling the matter directly. Why rock the boat, especially when efforts so often seem to fall on deaf ears? Why not bring it up next week instead of disturbing the dinner, or the vacation, or the marriage? Why not let it go entirely? It is safer and easier for a woman to keep quiet, to keep the peace, or to wring her hands and profess incompetence.

"Back in the late fifties and early sixties that was the way it was, and I tell you," maintains Wendy, her tone growing exasperated and her twang deeper, "it's exactly the same now, *exactly* the same."

I look around the young women at the University today and they have fluffed-up hair and little sweaters with teddy bears on them and they talk in little-girl voices and they're just trying to be pleasing and nonthreatening. And the girls that aren't trying to be nonthreatening aren't sought after. You act like this little piece of fluff that makes this boy feel like a big man, and he wants to keep on feeling like a big man but you don't want to keep feeling like a little piece of fluff.

A onetime fluffball herself, as the years went by Wendy realized there was no way to stretch her wings without brushing the walls of a very comfortable cage. Her privileged life was completely cir-

cumscribed. "Oh, you're Troy and Timmy Gorham's mom," the
lacrosse coach would say. "Oh, you're Dr. Gorham's wife, he operated on my dad," the grocery clerk would say. And no one could
imagine why a doctor's wife might want to work.

Once her sons were in high school and she had held every office
in the Junior League and the Women's Auxiliary, Wendy returned
to college for a master's in journalism. "Frank says my going back
to school ruined our marriage, and in a way he's *exactly* right,"
Wendy admits, slightly sheepishly, "because something had come
along in the interim called women's studies." Ambushed by the
women's movement, Wendy got caught up in the struggle for the
Equal Rights Amendment. Although "my friends were supposed to
be wives of his friends," Wendy's social circle expanded to include
divorced women and lesbians. Wendy began speaking out, both in
public and in private.

Frank never actually interfered with Wendy's new activities, but
they did put a strain on the marriage, as did her newly opinionated
self. "There is a contract in every marriage," explains Wendy,
"and the contract we got married under, and both bought into,
was that he was the doctor and I was the doctor's wife. He was the
wage-earner, I was the homemaker." That contract was changing,
and Frank didn't like it one bit. Wendy still did the laundry but it
stayed in the laundry basket to give her more time to write. "I
remember him parading through the house every morning with a
towel wrapped around his waist—we had a *big* house and it was a
long way from his shower to the laundry room level—and [he'd]
pick out one pair of socks and one pair of undershorts and go
stomping back. It became a lot of posturing," she recalls, suppress-
ing a giggle.

There was no going back, because what Wendy acquired during
those turbulent years was "a level of awareness, awareness of being
a fully franchised human being if you will. [I realized] that it was
all right for me to want an identity of my own." It took several
agonizing years, during which Wendy established herself as a journalist and cookbook writer, for it to become evident that the marriage could not contain a someone who was more than Dr.
Gorham's wife.

Toward the very end they decided to spend Christmas in Europe
with all three kids, and Wendy resolved to make it a great trip.

I kept my mouth shut and was just the sweet little doctor's wife, and the people on the tour just zeroed in on Frank the Doctor. "I have this pain ... ," "My friend has breast cancer ... ," he never minded that. At that time I'd sold a couple of books but I didn't say anything, my kids didn't say anything, Frank didn't say anything. And I was just invisible that whole time. There was something so incredible about backing off and looking at that. It was like my entire married life in capsulized form.

It turned out to be their last Christmas as husband and wife.

Like many women, Wendy spent several grueling years figuring out whether and how to extricate herself from her marriage. Although the high divorce rate implies otherwise to some, there is nothing casual about the decision to divorce. Both parties have an enormous amount invested in the status quo, and the choice between staying married and calling it quits is an agonizing one. Ending the defining, central, romantic partnership that marriage still represents for most of us is always wrenching. Shattering the illusion hurts. Even the most amicable divorce is still a wounding experience, and even the most wretched marriage needs to be mourned.

The end of a marriage is a loss, but not a failure. On the contrary it is a victory—over inertia, terror, conformity, insecurity, and countless other demons. Every woman who speaks in these pages has suffered enormously. Many wish they had left their marriages sooner, but had to wait until they had marshaled the financial or emotional resources. They are proud of themselves for having gone through with it, and not one regrets it. The process is a deeply personal one, and no outsider can determine how long it should take or what should be the breaking point.

CHAPTER 2

Hard Choices

A woman in a marriage gone wrong wrestles with the fact that she cannot stay and cannot leave, that both decisions have unendurable consequences. Where will she live? What will happen to the children—and how on earth could she do this to them? What kind of living can she make? Will she be alone for the rest of her life? The choice between her survival and that of her marriage terrifies.

The Lure of Limbo

"I can remember staring at the ceiling and saying, 'I don't know how I'm going to do this.'"

I know how tough it is to make the decision to end a marriage. Around the tenth year of my marriage, as discontent hardened into dull depression, the possibility that my marriage was ending entered my head, unbidden and unwelcome. A fight ended in my husband's shouting the D-word. I squelched the idea, banished it to the farthest recesses of my mental closets. Especially when I looked at my children, then four and six, I simply couldn't imagine rupturing their world. I clutched at straws. For a short, euphoric period I thought my marriage would be saved by the simple expedient of hiring a baby-sitter when I wanted to go out alone, instead of asking his permission. Of course it didn't work, and the three couples therapists we visited over the course of that year couldn't save the marriage either.

I was shocked that a wave of relief, not panic, swept over me when we agreed to a trial separation and my husband moved out. Six weeks later, he announced that he had worked through our problems and was moving back in. At this point I told him that I wanted a divorce. I felt enormous rage and helplessness when he moved back in, and for seven months we lived again under the same roof with our two children. That period was scarier and more disorienting than I could have imagined. I was a wreck. I stopped eating. I'd lie awake at night wondering if I was wrecking my kids' lives, if men would still look at me, if I'd end up in some dump stretching the meatloaf with Hamburger Helper. None of these fears came true, but they were very real at the time, and they have a very long half-life.

I learned two important things from this experience. The first is that it's the period before the decision—waiting, wavering, calculating, tucking in blissfully ignorant children (far less ignorant than parents think), anticipating the unknowable—that is the grimmest. Just as when I was pregnant for the first time, all I knew for sure was what was being given up. Once the decision was irrevocable, the sure knowledge that I was doing the right thing for all of us never left me. "The limbo was the worst," agrees Virginia, a thirty-nine-year-old cartographer, "but when I finally decided it was over, I felt very powerful. It was scary as fucking hell, but there was tremendous clarity, which was a relief after three years of wondering."

The second surprising but immensely comforting fact emerged in the course of researching this book: only one of the women I interviewed regretted marrying, and not one was sorry she had divorced. The only regret shared by almost all the respondents was that they had stayed married as long as they did. An acerbic Yolanda sums it up: "my own stupidity and infantilism were the only things that stood in my way. I thought I was born without the gene for car insurance. I should have left earlier."

The Power of Fear

Many women, when asked what emotion predominated at the time of the decision to end the marriage, said fear. Relief came in a very close second, however (with anger a distant third), and many women seesawed back and forth between the two. It might seem that fear and relief would cancel each other out, but their coexistence makes sense. It reflects the fact that the fears being faced are complex, paradoxical, even contradictory. Some are rational ("I won't have as much money"), others irrational ("I'll never be able to make it on my own"). Some are legitimate ("I need to learn how to cope without a man in my life"), others less so ("No one will ever love me again").

The tendency is to see the future in terms of evers and nevers, blacks and whites instead of shades of gray. In the movie *Sleepless in Seattle*, Rosie O'Donnell's character says she's heard that a woman over thirty is less likely to get married than to be shot by a terrorist. "That's not true," protests Meg Ryan's character. "Yeah," counters O'Donnell, "but it feels true." The truth almost doesn't matter; that's the way we *feel*, and feelings are what resonate.

Especially during periods of transition, doubts and insecurities surface with a vengeance. One fundamental response is to do nothing at all, like the proverbial deer frozen in the headlights. Fear can paralyze. It's also hard for a depressed person to take initiative. For reasons ranging from terror to torpor, the women in this study spent months or years wrestling with whether and when to make the move. Though it seemed easier at the time to sit tight, this immobility clearly took a tremendous emotional toll.

In *Unfinished Business*, her book on depression, Maggie Scarf observed a tendency toward stillness and inaction in women overwhelmed by fear of the unknown. "But this quiescence, from the point of view of outward behavior, was being experienced with a terrible inward violence," writes Scarf, going on to note that one factor more than any other triggers depression in women, "and that is the loss of a love-bond." Eyes open slowly. The mind is not eager to process what it sees. Immobility is more comfortable than the prospect of divorce until fears are faced and put in perspective.

FEAR OF AUTONOMY

Despite the fact that she made a good living and had no children, Jodie, a slightly built software analyst from San Francisco whose auburn curls frame her narrow face, spent more than half of her four-year marriage in an agony of indecision. She'd waited until she was thirty-five to marry so she'd be sure to get it right; how could she have gotten it so wrong? Paul was the child of alcoholic parents, and his need for absolute control over his environment came as a rude shock. He insisted that Jodie take his name, cook macrobiotically or not at all, cater to his unorthodox sexual tastes, and be on call to respond immediately to his needs. Within a year, she knew she'd made a mistake. "It was like he wanted me to bring my gregariousness and happiness into his life, and then to channel and control it," she says, still sounding bewildered. For example, if she was on the phone when he came home, he'd interrupt until she hung up. It wasn't because he wanted to talk to her; he just didn't want her to be on the phone.

Unable to schmooze, to have people over, to buy as much as an ice tray without Paul's approval, Jodie eventually gave up "any sense that I could be empowered in any way." She and Paul worked in the same field, and despite an impressive premarital résumé, Jodie began to fear that friends and co-workers thought well of her only because she was Paul's wife. She was afraid that in a divorce her parents would support Paul, the son they'd never had, instead of her. She thought people would think she was nuts or somehow defective for not hanging on to this attractive, successful guy.

The main thing that struck her was "how much I wanted to get divorced, how destructive to me that relationship was, and yet how hard it was to let it go, so goddamned hard." Instead of seeing what was happening to her, she worried that Paul, whose mother had died when he was young, would feel she was abandoning him. Some masochism was at work here, but Jodie was also imprisoned by the premise that underlies traditional marriage: that a wife's desires coincide with those of her mate and children. "The terror is that if she responds to her own needs and wants, everything else will go to hell; that only women's self-sacrifice assures reliable family bonds; her desire promises chaos," writes Dalma Heyn. (Heyn notes that this does not apply to men, who have long made time for golf or guerrilla warfare without jeopardizing their families' welfare.)

Pushing forty, Jodie was terrified of changing absolutely everything in her life, despite acknowledging that "I didn't even know who I was anymore." Vulnerable to start with, Jodie's personality was being obliterated. Would she be alone forever? Would it better to be alone and unhappy than unhappily paired? Jodie had been in therapy for years, but had only recently started to address her own inertia. It was her therapist who finally told her bluntly, "You have to take action, to start living again." There are tears in her eyes as she recalls finally confronting what had happened to her, and the prospect of coming full circle.

> I had always looked after my own stuff financially, I had always done everything. And I gradually surrendered, until I felt so numb and so not-alive and so controlled and so without voice that I thought, "Oh my God. Part of what's really rich and wonderful about life is that you *get* to take care of yourself. It's a privilege: you *get* to do it."

She spent a month hiking in Thailand, accepted a job in a different city, and called a lawyer. For Jodie deciding was the hard part, not carrying through with the decision and dealing with the consequences.

The consequences *are* painful and scary. The grief and loss are real, and anyone going through a divorce passes through a whole trajectory of wrenching emotions. (This process is discussed in detail in Chapter Five.) For Eileen, the prospect of taking off her rings was especially painful. Prim and proper in her pressed slacks and white turtleneck, she winces as she recalls, "At the supermarket I had this feeling that everybody would go, 'Look at her hand. What's wrong with her?'" The fear wasn't rational, of course; Eileen soon realized that no one but her close friends would notice, but the symbolic gesture was still difficult.

FEAR OF FINANCIAL HARDSHIP

Perhaps the most universal fear divorcing women face is financial, and it is true that divorce drastically reduces the standard of living for many women. The current consensus is that divorced women suffer a 30 percent loss in income. One-third of families headed by women live below the poverty line, which was $13,359 for a family of four in 1990. (The issues behind these figures, and how they play out in the lives of divorced women, are discussed in Chapter Four.) But even the neediest women I interviewed felt their overall situation improved after divorce. Hannah, "petrified" when she took her four children away from their Wisconsin farmhouse, had to go on welfare for a year and a half. She nevertheless felt better off away from her controlling husband. "It's amazing how, even if you're poor, if you have some sense of choice and control over what you do and whatever you have, you can feel quite well off," she says, the point clinched by her proud, shy smile.

When married, Hannah handled no money except the minimal amount Ed would dole out for trips to the grocery store. Hannah didn't aspire to work outside the home; being a mother and home-schooling her four children was her job. Unfortunately it had no value in her husband's eyes, and fundamentalist Christian beliefs reinforced his authoritarianism. After the divorce Hannah and the four children lived on $679 a month from ADC, making ends meet thanks to the cheap rental of a friend's house. (Intended to help indigent widows keep their children out of orphanages, Aid to Dependent Children was instituted as part of the 1935 Social Security Act. It became Aid to Families with Dependent Children—AFDC, generally referred to as "welfare" though still called ADC in some states—in 1962.) Hannah feels her standard of living improved. "I don't think there actually was more money, but having the control over it meant that for me, in reality, it was more." She laughs softly. "The kids and I could go to a store and buy whatever groceries we want with the food stamps, we had everything that we needed." While she was on ADC, the $90 a week Ed paid (or failed to) in child support went directly to the state. Hannah informed the court when Ed got a full-time teaching job so his child-support payment went up to $145 a week, which is now taken out of his wages and paid to Hannah directly. "We do okay," is her assessment.

During that first year a minister called to offer Hannah and her family assistance. He said, "You know, our church has funds to help, we know you've just gone through a divorce, that you're alone with the kids, if there's anything we can do, financially or otherwise . . ." Hannah replied, "I really appreciate that, but I have to tell you, I feel very rich." Hannah scrambled for various part-time jobs, including typing medical reports for a doctor in town. When she and her children moved in with him a year and a half later, the first thing she did was call up ADC and say, "'I want off! I don't need you anymore!' It felt *so* good," says Hannah jubilantly. She and the doctor have gotten married, but that doesn't mean she plans to stop working. "Mitchell earns enough money that I don't really need to, but I *like* it. I don't ever want to be dependent on somebody again. This fall I want to go back to college, in psychology."

"I'm in debt. That's what the divorce has done to my standard of living," says Nancy matter-of-factly. She is one of a number of women who worked full-time at good jobs—a tableware designer, she works for an interior-design partnership—but who earned significantly less than their husbands. Eric greatly outearned his wife as an advertising account executive; their combined income was $100,000. Two years after the divorce she has worked her way up to an income of $50,000. One reason it's not a hardship is because she was used to cutting corners, because "Eric was always screaming poor, saving to set up his own business." Nancy feels that her standard of living is actually better post-divorce, in the sense that "I know what I'm controlling. For me that's the comfort level," she explains. "We make sacrifices, but I don't feel like I'm missing out on stuff. My daughter and I don't do a lot of things a lot of people do, but we're perfectly happy." With evident relief and pleasure, Nancy adds, "I know what I'm walking into now. I know that this is trouble, or I know what my debts are, but I *know*."

For woman after woman the pleasures of self-sufficiency overwhelmed material loss. Once again, their *feelings* about their circumstances—whether rational or not—anchored their experiences.

FEAR OF REJECTION BY LOVED ONES

Another deep-seated fear is losing the support and approval of friends or family. Sometimes it's well-founded. For example, when a woman has an affair before ending her marriage, her friends and family may perceive her as the bad guy. In the early seventies, as the divorce rate dramatically increased, many women were shunned by neighbors as if they were the carriers of a social disease. Monica got divorced twenty years ago, at age thirty-seven, one of the first women to do so in her posh Winnetka, Illinois, suburb in what she calls "the Pleistocene era" (1975). Sporting funky round glasses and polka-dot tights, this fashion consultant still sounds bitter when she talks about the woman who "took it upon herself to imagine that her husband had the hots for me," in Monica's words, and turned her back on her best friend.

For Monica, that betrayal was more horrendous than that of her family, who predictably looked the other way. Nor did she get much help from her Winnetka neighbors. "People would stop me in the supermarket, neighbors, and they would say with furrowed brows, 'Are you sure you know what you're doing? What about your son?' In fairness, lo these twenty-one years later," Monica adds, "I could see fear in their eyes, intimations of what could happen to them if their husbands suddenly took off." Indeed, ten years later, some of those same women were calling Monica up for advice on how to get through their own divorces.

Today many women still end up with neighbors and families who are deeply disturbed or angered by their divorces. Delores has been ostracized from her family and the traditional Cuban society in which she grew up. "You're talking to an old-fashioned girl here," she says. "In my culture, the way women are trained is you marry, you die, you make bones with the husband. Good or bad, he's your husband." Now thirty-three, she could still pass for a schoolgirl; a long black braid runs down her back, her flowered blouse is buttoned to the neck, and her sneakered feet are demurely crossed at the ankles.

Medieval though it may seem, Delores's was an arranged marriage. Carlos knew he had it made in the shade, and when Delores announced she was leaving with their one-year-old son, Miguel, he thought she was kidding. He said, "You can't do that, Delores. You know how bad being a divorced woman is?" Her parents made it clear that if she dirtied their name, she could count on no help from them. They kept their word and turned their backs. A shy and private person whose son is the center of her life, Delores is profoundly isolated, and struggling financially because Carlos pays no child support. But though she deplores the fact that it was necessary, she considers the divorce "the best thing I ever did." She's finally in control of her life, she has her faith, she has

Miguel, and she has no regrets. "None at all. I'm happy I did it."

Some friends and even family members do fall by the wayside. But as with other kinds of upheaval, divorce strengthens some bonds and creates new ones. Countless parents took their daughters in until they could get back on their feet, and many women, especially those with young children and without other resources, said they couldn't have made it without their families' help, whether it was a brother picking up belongings in his pickup, an aunt looking after the baby, or a sister lending money.

Many women find that once word is out that the marriage is over, their friends and family quickly reveal that they had never liked the man all that much and are delighted to have him out of the picture. Still, love and loyalty don't evaporate overnight, and it can be hard to hear one's ex denigrated. I well remember breaking the news to my parents that I thought my marriage was in deep trouble, and my next sentence: "I don't want to hear a bad word about him," a request they honored. Finger-wagging is unwelcome and hindsight gives an unfair advantage, but unconditional support is a blessing.

A nice paradox was that in the process of becoming self-reliant, some women learned whom they could rely on. Sometimes support came from an unlikely source. One woman found that a sister-in-law was the only family member who genuinely sympathized. Another called on two childhood friends rather than on her family, who felt rebuffed that she didn't want to move back home at age forty-two. Beverley, married to a minister, ended up estranged from the community she had grown up in but found herself a new church with a great women's ministry. For some women, divorce is a lesson in reaching out. Especially for those who prided themselves on their self-reliance, the process offered wonderful surprises and an unexpected wealth of love and support.

Some women have the confidence to write off their detractors, figuring that if certain friends can't be supportive, well, it's their loss. Of course, not all divorcing women can muster such disregard for other people's opinions. Many are extremely apprehensive about the outside world's perceptions of them, concerns that in retrospect may seem overblown, even paranoid. But given the current pro-marriage, anti-divorce political climate, guilty feelings are understandable.

LORRAINE

I wish I hadn't spent so much time trying to make him feel good about something he wasn't destined to feel good about. He kept coming to me and saying, "*Please*, let's try again, let's do this one more time, why can't you, why can't you . . ." It probably would have made as much sense if I'd just said, "Good-bye. I'm outta here." It would have been easier on him in a way, and it would certainly have been easier on me.

The Power of Guilt

The conservative party line is clear: the virtuous stay married at all costs. Those who leave their marriages are guilty of outrageous selfishness, except in the case of flagrant infidelity or physical abuse. Just as much damage is done to the union when the more subjective elements of the marriage vow—to cherish and honor—are violated, but these needs are perceived as "softer," less important. The good wife, conditioned to believe that her needs are selfish and insignificant, dutifully stays put.

Always marriage-happy, American popular culture has gone pro-nuptial with a vengeance. Movies featuring independent women, like *An Unmarried Woman*, would never be made today. Instead, stand-by-your-man movies are back in vogue, such as *Leaving Las Vegas* (even a suicidal, unfaithful, alcoholic is better than no man at all), *How to Make an American Quilt* (don't quibble if hubby wants to reserve that extra room for a nursery rather than an office for his wife), and *Bridges of Madison County* (four days of bliss is a fair tradeoff for umpteen stultifying years of marriage). The writer of the controversial bad-girl movie *Thelma and Louise* followed up with *Something to Talk About*, an homage to working it out in the face of infidelity and insult. In the 1970s and 1980s *Men Behaving Badly* would have been the title of a serious drama about the difficulty of being a single woman; in 1997 it's a sitcom about zany-but-lovable, can't-live-without-'em fellas. Men may be from Mars, but now they're admitting that Venusian-free life is lonely. The battle of the sexes has given way to a truce presided over by the gurus of improved communication skills.

These messages are reinforced by energetic media coverage of the negative aspects of single life, separation, and divorce (as exhaustively documented in *Backlash*). Even when they are tenuously researched, reports that divorce devastates children and impoverishes and isolates women make headlines; rebuttals do not. Scary stories sell, especially those that play into our worst fears. Everything from illiteracy to the murder rate is blamed on single-parent homes and the reputed ease of getting divorced. "When you ... make divorce quick and easy, you don't need a Ph.D. to know what will happen," declares David Blankenhorn of the Institute for American Values. "You'll erode the American family." This reasoning is circuitous and simplistic. The American family is being eroded by the factors that *lead* to divorce. Nevertheless these myths about the devastating effects of divorce are potent, inflicting massive guilt upon every woman who contemplates leaving her marriage.

"I'LL WAIT UNTIL SOL DIES. NO, I'LL TRY TO KILL SOL. NO, I'LL KILL MYSELF ... "

Even women like Esther, a retired bookkeeper whose children were grown at the time she moved out, are considered selfish by the culture at large. Unaccustomed to putting their needs first, they can be immobilized by guilt. The prospect of telling her elderly parents that she wanted to leave her husband stopped Esther in her tracks, even though watching them slide into an unhappy old age together was one of the reasons she herself wanted out. Now sixty-nine, with unruly gray hair pinned back in a plain metal barrette, she's clearly comfortable with the single life she's made for herself after a quarter century of married life. Fourteen years earlier, moving into her fifties, she looked ahead and found the prospect of growing old with her husband, Sol, boring and saddening. "I see marriages like mine in the supermarket every day," she says, ever the realist. "He didn't drink, he didn't beat me, he didn't gamble. To me it was a very ordinary marriage, but I wanted out of that ordinary marriage."

The guilt over this decision immobilized Esther for many months. "One of my thoughts was, 'I can't tell my parents. I'll wait until they die. No, I'll wait until Sol dies. No, I'll try to kill Sol. No, I'll kill myself ... '" recalls Esther with a wry smile. She decided she and Sol should see a counselor, but at the last minute he changed his mind—"I hear a lot of women saying their husbands say the same thing: 'I don't need it, you need it, so you go,'" Esther comments—so she went alone. When she called Sol to come pick her up afterward, the counselor remarked, "You still need him." Says Esther, "This was like a lightning bolt. I could have called a cab!" It took a year in counseling to figure out that waiting for her parents to die was absurd, and that getting divorced was something she had to do for herself.

Full of dread nonetheless, she drove over to her parents' apartment, screwed up her nerve, and told her father, "I'm leaving Sol. This is it." Her father, a building superintendent, responded in the most reassuring way he knew, by offering to get his daughter an apartment in the building. Esther's mother said simply, "'I knew you weren't happy.' They were both supportive. It surprised the hell out of me," admits Esther.

Though her parents were unexpectedly accepting, Esther still had to deal with her husband, who went to pieces once he realized she was serious. Sol played every guilt-inducing card he could think of, crying on the kids' shoulders, calling Esther up to say, "I don't have any hot water, can I come over?" and even threatening suicide. To Esther's relief, largely because of the strain on the three kids, Sol remarried within a year. "He was threatening to throw himself out the window and it turned out he'd already met somebody," says Esther with a laugh, and without an ounce of envy. She remembers consulting a divorced relative at the time, who described "opening the door to her new apartment to complete quiet, like a little heaven on earth." Esther feels the same way about her modest one-bedroom in West Palm Beach. Guilt is a distant memory.

BENDING OVER BACKWARDS

One way women attempt to overcome their guilt at wanting out is by trying in every way imaginable to keep their marriages together. Efforts range from enduring abuse to attempting endless reconciliations, from dieting to putting up $30,000 worth of designer wallpaper on the walls of a new house. "That was $30,000 that I could have used," sighs Keiko, a real-estate agent in Seattle, knowing how ridiculous it sounds, "but it was a gesture of commitment. A silly one, but a gesture of commitment." Theresa, a nurse from Naples, Florida, went to a homeopath for herbs that were supposed to increase her sex drive, until "I realized I'd rather screw a doorknob than this guy," recalls Theresa, her green eyes shining with laughter at the absurdity of the mission.

The well-meaning and often ill-considered actions of these women belie the guilt-inducing message of the religious right that divorce is the facile refuge of the self-indulgent. "I'm the one who did it, I divorced. I pulled the moral fabric of this country apart. Selfish quitter," declares writer Ann Patchett, describing how she felt when she left her husband. Her feelings were real, but Patchett is also poking fun at the notion that all of America's ills can be attributed to the breakdown of the family, and by extension to the person who initiates divorce. It's a mistake to assume that the spouse who asks for the divorce is the one who made that step inevitable. It's also wrong to assume that the decision is casual or impulsive; that was not the case for Patchett or any of the women interviewed in this book, who agonized for months and even years about big and ultimately unresolvable issues.

WHO SAYS IT'S CASUAL?

If moral sloth is at the root of the high divorce rate, as conservatives contend, why do women work so hard to save their marriages and so readily assume that the burden is theirs? Marriage is hard work, but more so for women, who are expected to accommodate and adapt because, after all, boys will be boys. That's what underlies the decision, almost always the wife's, to seek counseling. Women, always willing to contemplate mental and physical overhaul, constitute the great majority of psychological patients and readers of self-help books.

Each woman in this book struggled long and hard to meet a central challenge: to connect with her husband and support her marriage without losing herself in the process. For example, Nancy, the tableware designer, spent two years struggling with how to make her marriage work. She went to counseling, though her husband refused to accompany her because he figured that resolving any problems in their marriage was her department. "This guy will never have anyone in his life think about him, and try and be fair about him, and try to understand him as much as I did in that two-year period," Nancy says harshly. "He does not know what he has thrown away. And he never got it. He *never* got it." Nancy describes flashes of deep, deep sadness during the process and also afterward, but she went out and celebrated when the papers were signed. "I was absolutely delighted, absolutely delighted. I guess because in my heart of hearts I had stayed for two years making this decision. It was *not* an easy decision. But once I decided to divorce, it was very clear that it was right, it was full steam ahead."

Nancy's years of struggle to save her marriage, far from unique, also belie the sanctimonious promise of weekend seminars that any marriage can be saved by two days of fine-tuning relationship skills. Christian columnist Michael McManus leads a Quebec-based program called Retrouvaille, which claims to patch up the marriages of 90 percent of the five thousand couples who participate in its seminars each year. "In the Retrouvaille world no marriage is too broken," wrote Hanna Rosin in *The New Republic*, observing that McManus has rescued the unions of alcoholics, abusers, and adulterers. Another conservative concerned about the breakdown of the American family, Bill McCartney in 1990 founded Promise Keepers, a group that embraces the biblical model of how husbands should behave. Men flock to stadium rallies to chant "Thank God I'm a man" and embrace the message that if they just "take back" power from women and take over all the decision-making, their marriages will succeed. Despite all the good intentions in the world, soul-searching and better communication sometimes reveal that two people shouldn't have married in the first place. Not all differences between husband and wife are reconcilable, nor should they be.

Getting Unstuck
"It was hard, the hardest thing I've ever done."

ROSEMARY

I would say spend at least six months doing your homework before you even talk about it, figuring out whether you really want to do it, figuring out how, testing the waters. Just really paying attention to what you're doing and whether that's what you want. I have not been sorry one day, not one day.

In retrospect, many of the women in this study said their number one regret was not getting out earlier. Some toughed out the transition on their own. The majority did seek counseling to sort out feelings and issues, and most found it invaluable. Without it, some women said they probably would never have had the strength or clarity to end their marriages.

Individual Counseling

A counselor, or therapist, helps a troubled person see issues more clearly: whether she is blaming the relationship for problems that won't end when it does, for example, or how to interpret the reactions of children of different ages, or how to evaluate the short- and long-term consequences of different courses of action. A counselor is objective in a way a friend cannot be. The counselor's job is to help you to change your life, or to become more comfortable with the decisions you've made. A good therapist doesn't supply answers, but guides the patient toward honest self-appraisal and the difficult process of change. Growth is painful, and the responsibility is the patient's, on the couch as in life.

Much to my irritation, my therapist never coughed up any of the direct advice I blatantly fished for. I began individual therapy two years before the divorce, and ended more or less when the papers were signed. Having inherited my mother's distaste for introspection, I didn't enjoy it, but my therapist's guidance was crucial. She helped me see my part of the patterns in my marriage, and, more importantly, to glimpse who I would be alone when I was no longer married. I'm sure I would have gotten divorced eventually, but therapy helped me do it sooner and better. As one woman I interviewed, Celeste, said: "It was my growing unhappiness that drove me into therapy, and my growing consciousness that made me end the marriage." Similarly, Megan, the cable-TV salesperson from Boston, went to a counselor for two full years before she and Anthony separated, figuring that "maybe if I got my head fixed I would feel better about the marriage." Instead she started to feel better about herself and realized she wanted out.

Susannah, a poised preppie whose long chestnut hair is held off her face by a velvet headband, married a much older artist whose imagined superiority whittled away much of her self-assurance over the six years they were together. Before actually moving out, she decided she should talk to someone to make sure she was doing the right thing, and a brief stint of counseling was enough to provide her with a new perspective on herself as well as on her marriage. Susannah picked a shrink who knew her husband Stan and their whole circle of friends, and who promptly turned the issue around. "I think the key question here is not why you're leaving him," he said. "The important thing is to find out why you married him, because you don't want to do that again." Now a film archivist in Berkeley, Susannah remarried fourteen contented years ago. "I certainly am different from the way I was in my first marriage, when I was a pushover," she says. "That was the whooping cough, and I'm not getting that again. That was the end of childhood."

Nancy, the tableware designer, wisely chose a therapist who she felt would challenge her to grow and face her own failings. She actually switched to a male therapist during the divorce, because the woman she had been consulting "ended up kind of male-bashing, and that wasn't what I was trying to do. There are two sides to this. You can't go around saying, 'I'm perfect; this person did this to me.' You do it to each other." Caught up in the minutiae and decision-making of the divorce process, Nancy "needed to be in therapy to have someone pull me out and say, 'You've got to look at the bigger picture.' It really helped me through that, in kind of getting a sense of myself." The change is always painful and sometimes unwelcome, but the reward—that elusive sense of self—is worth it.

When just getting through the day seems heroic and the smallest move means shouldering a ton of emotional baggage, seeing the big picture seems nearly impossible. That's where therapy can be invaluable. Though it may not save the marriage, it may rescue the people involved.

Marriage Counseling

Perhaps because consulting them is often a matter of too little too late, a common sentiment is that marriage counselors serve largely to make it clear that a relationship is hopelessly derailed.

Though far from their only function, this is a legitimate one. In some cases the purpose of counseling is not to bring about reconciliation but to smooth out the mechanisms of separation. Over the course of our last year and a half together, my husband and I saw three different counselors. In retrospect the visits served largely to confirm our irreconcilable differences, but it was comforting for me to have a neutral place to discuss loaded issues. It salved my conscience too; I wanted to feel that we had done all we could.

Picking a counselor that suits both spouses is tricky, since each, inevitably, is seeking an advocate. Some men are not comfortable with a female therapist. I ran into this problem with the pull-no-punches shrink I nicknamed the Israeli Commando; I thought she was great but my husband pulled the plug after three visits. Many husbands are reluctant to seek counseling, refusing outright or backing out at the last moment. Anneke's husband went the other way. Every time the romance writer brought up divorce, he figured she was going through some sort of phase and he would say, "We'll go back to therapy." Eventually Anneke had to refuse.

Some people, for whom therapy still carries a stigma, feel embarrassed or ashamed about seeking help. When Marjorie's husband refused to acknowledge the existence of any problems, marital or personal, Marjorie decided to seek help for herself. This was an especially difficult decision given her Chinese-American upbringing, though she had been educated in the States. "The Chinese mind-set about going to see a psychiatrist is that, 'Oh, you're really, really sick.' It's very taboo," she explains. Marjorie had a real block about it, but she knew she needed help because "for some reason I literally couldn't spit that word [divorce] out of my mouth." Individual therapy got her past that roadblock, but it was in marriage counseling that Marjorie really made up her mind. "As we discussed the history of our marriage, the two of us had *totally* different memories of every incident," she recalls, a bemused look coming over her plump countenance. "I looked at him, he looked at me, we were astounded. I said, 'This just shows you how different we are.' So one day when we walked out of the therapist's office, I said, 'Let's do it.'" This phenomenon, very different memories of the same event, is unsurprisingly common among couples who divorce.

SEEKING PROFESSIONAL HELP

Research shows that people find treatment at the hands of psychologists, psychiatrists, and social workers equally effective. A psychiatrist is a medical doctor, who can prescribe medication. Psychologists have more initial training than clinical social workers, but a social worker with experience may be just as helpful.

Fees vary considerably. Go with what you can afford, and check your insurance plan; most companies offer at least limited coverage. Look for a therapist who is licensed (requirements vary state to state), and avoid practitioners with little or no formal training, or titles you don't recognize.

Most important is to find a therapist you feel comfortable with and one who is respectful. Some women feel more comfortable with a female counselor; others want a male point of view. Whoever you choose, don't hesitate to inquire about his or her training and expertise, and don't hesitate to change therapists if it turns out not to be a good fit. Many issues can be addressed in six to twelve sessions, but other problems require long-term treatment.

There are many ways to find a therapist, starting with referrals from a doctor or friend. Your place of employment or trade union may offer counseling services, as do most universities. Look up "Mental Health," "Therapists," "Psychologists," or "Psychotherapists" in your local phone book, or ask the national associations of any of those professions for the names of several therapists who specialize in your area of concern and live in your community. The phone number 800-THERAPIST will connect you to a nationwide referral system of mental health professionals and also provide general information about different mental health disciplines, and a number of directories of therapists are available on the Internet, such as http://www.psychology.com/therapy.htm. and http://www.counselorlink.com/.

Another option is one of the 12-Step programs such as Al-Anon or Adult Children of Alcoholics, which are self-help recovery programs for families and friends of alcoholics, or Codependents Anonymous, a similar program for people in dysfunctional relationships who need to focus on their own needs. Other 12-Step programs address a variety of behavioral problems such as depression and mood disorders, overeating, drug and alcohol abuse, and other addictive or compulsive disorders. Meetings can be found nationwide; consult your local phone directory. Membership in 12-Step programs is free, and anonymous. Members admit they need help, acknowledge a higher power, and receive guidance, fellowship, and support in the company of equals.

Al-Anon and Other 12-Step Programs

For those without the resources to pay for therapy, Al-Anon or other peer counseling groups can be the answer. For Helen, a heavyset woman with an open, homely face who spent her married life in a rusty double-wide trailer outside Tallahassee, the price was right and the counsel superb. "Al-Anon saved my life," she declares. In the first place it woke her up to the belittling and codependent dynamics of her marriage, of which she had been completely unaware. Secondly, "Al-Anon helped me find the courage to make the change, to walk away from the marriage and not look back."

The 12-Step programs can also serve as an adjunct to therapy. Although Eileen, a prim and collected hospital administrator, was also going to a counselor three times a week, she gives Al-Anon the credit for getting her through. Sometimes, when things were really bad, she went twice a day. "At Al-Anon, see, it's this constant reinforcement, 'Yes, you can do it. Any problem can be solved. There is a solution. You will come to it,'" she explains. For her, the program's major benefit "was to give me the ability to not react to everything, to be able to step back, to really see what was going on, and to be consistent myself. I'm still the same person I was ten minutes ago. Even if you're screaming at me, I'm still the same person." She needed that strength to wean herself, and her manic-depressive husband, from codependence. Eileen grew up in an alcoholic family and wishes she had found Al-Anon sooner, in time to help her through childhood and into her adult years.

Leaving Abusive Marriages

"You hear that negative stuff long enough, you believe it."

Because abuse takes so many forms, and because the word has very different connotations for different people, it is necessarily a relative term. There is a difference between being attacked with words and being attacked with fists, though words can be just as emotionally destructive. Lines blur. What abusive marriages have in common is that such behavior is rationalized and minimized by both partners. A hellish but familiar present is preferable to an unknown future. For some the turning point is when the abuse turns physical; for some it's when the safety of a child is compromised; and for others it's an epiphany that the situation is never going to improve, a less predictable moment of truth. Whatever the particulars, the element of abuse and a wife's complicity in it vastly complicate the dynamics of deciding to divorce.

For Helen the turning point in her marriage came one night when an argument ended in Ron hitting her, "a good old hay-baler right across the side of the face," and Helen made up her mind not to become a statistic. She had stayed put for ten unhappy and fearful years, during which alcohol turned a childlike and irresponsible man into a withdrawn and abusive one. "I had to watch what I said, I had to watch what I did, because I never knew what was going to set him off," she remembers, wincing at the memory. "I could look out the window of our trailer on a Friday afternoon when I'd hear his car pull in the driveway, watch him get out of the car, and I would know immediately what kind of a weekend I was going to have."

Susan Brownmiller notes that "'women's intuition' . . . may be nothing more than defensive watchfulness, a picking up and putting together of verbal and nonverbal cues as a strategy of survival, as the subordinate animal is sensitive to the sounds and movements of the dominant animal, which does not need to think before it acts." Helen's behavior on Friday afternoons was a perfect example of this adaptive behavior, common to many households where the husband's mood sets the tenor of the outing, the mood of the party, the tone of the evening. Whether that makes these marriages abusive is often unclear even to the participants. Each person has a level of tolerance known only to herself, and a breaking point that no outsider can predict or impose on her.

Substance abuse plays a part in many destructive marriages. Addictive personalities may be seduced by more than one substance. Different drugs affect people differently, influencing emotions, mood, and behavior. Their use can be both a symptom and a cause of problems, and even moderate use can cause trouble. Substance abuse can worsen preexisting problems, leading to estrangement between spouses who partake and spouses who don't, to financial irresponsibility, and to violence.

Alcohol and Other Substance Abuse

Part of the difficulty in dealing with a spouse who drinks is that not everyone who drinks is an alcoholic. Drinking may also be paired with other substance abuse, and separating out the issues can be challenging. Ginny's husband, Dan, used drugs and was immature and financially irresponsible, but his alcohol abuse was the clincher. Ginny realized her marriage was doomed when despite AA, counseling, and countless interventions, her family was still being held hostage to Dan's drinking. After two pregnancies and twenty-five years, Ginny's figure is still that of the teenager who won two statewide figure-skating championships. There's a raw-boned intensity in the way she tosses her words and hands around, and she describes herself as the "violent, loud one" in contrast to the well-off, happy-go-lucky guy whose parents ran a ski resort near Lake Tahoe and whom she married at twenty-four.

In the beginning Ginny and Dan were very happy. Then came parenthood, twins in fact. "He was used to a kind of party atmosphere all the time, and suddenly we had these kids, not just one, which you could sort of pass off, but two," she points out, "and it was just completely overwhelming." Dan started drinking pretty heavily, and generally steering clear of responsibility. Over the next few years, largely unbeknownst to his wife, Dan started snorting cocaine, drinking more, and spending a lot of money on things like big boats.

When the twins were four, Ginny got pregnant again. When she refused to consider abortion, Dan flipped out, and the night she came home from the hospital, he went out on a three-day binge. Ginny recalls it with a shudder, a freckled hand impatiently pushing a wayward strand of strawberry-blond hair out of her eyes.

It was just awful. To this day—and I have gone through years of therapy and Al-Anon to try—I just can't overcome that one thing. A counselor finally said, "Maybe you're just not supposed to forgive him for that." Because I will never, ever be able to get over that feeling. Here I come home with this *gorgeous* child, and you go out and pretend it hasn't happened? I kept trying to do more, get a job, do this and that, but no matter what I did it didn't make it right. Until finally I knew that I could not live one more day, not *one* day, in this house with this man.

"Finally" was after fifteen years of marriage, when that baby, now an eight-year-old-girl, smelled alcohol on Dan's breath. Drawn to the hidden bottle of Budweiser as if by a magnet, Ginny came up the basement stairs and threw it in her porcelain sink. "It never broke. It was one of the miracles in my life, and I truly believe in miracles, because I was going to kill him with the broken bottle," she declares flatly. She told him she wanted him out that very night, and he left. At the time she thought it was going to be for three or four days, but on the very first day "in our house there was a real calm. I knew I would not give that calm up to have him back."

Verbal Abuse

Ginny never faced verbal abuse from her spouse, but many women do. They are immensely susceptible to the corrosive sound-track that plays through their marriages, believing themselves in turn bad, ugly, dumb, undesirable, sneaky, helpless, or incompetent, despite abundant evidence to the contrary. Called "stupid" or "crazy," women hasten to back down when confronted. They look around at their marital beds, their sleeping children, their Tupperware, and their deeply familiar domestic landscapes and figure, yes, they must be crazy to want something different.

Verbal abuse can run the gamut from the occasional wounding remark to systematic derogation. A prime factor in one woman's divorce was the shocking realization that she was using so much energy to block out her husband's constant harassing and humiliating comments that she was actually losing her memory. Accusations may be grossly exaggerated or downright irrational, but this doesn't necessarily make any difference to either party. Women in abusive marriages are frequently better educated than their husbands, as was the woman who described going "from a very intelligent woman to a complete idiot, about everything," but such advantages provide little protection. Helen was not alone in observing that "you hear that negative stuff long enough, you believe it." Women don't necessarily call it emotional abuse, a term that smacks of victimspeak, but that's what it is. Many husbands are routinely derogatory. It makes them feel superior, and it makes their wives more dependent on them. Women internalize the criticism, not just because of the tendency to blame themselves for problems, but also because it hurts to admit that someone you love is just plain mean.

Forty-nine years old, Susannah still has the thick chestnut hair and bouncy stride of the confident valedictorian of her boarding school class. Although she describes her ex-husband as a "very sweet person" who praised her effusively to friends, when alone with her Stan dished out "a very bizarre, clever kind of verbal abuse" to which she hadn't the foggiest idea how to respond. He was passive-aggressive and extremely clever, and knew exactly how to push her buttons. "I'd get a bad haircut and he'd say, 'You look like a cross between Mick Jagger and a faggoty newscaster,' or 'You're looking like a Scarsdale matron,' or 'That looks like a copy of a coat Jackie Kennedy would have looked at but not bought,'" recalls Susannah with a wince.

The film archivist, who now lives in Berkeley, was the second of three daughters of a high-powered chemical-company executive. In what she characterizes as "the most pathetic rebellion" from her traditional upbringing, at twenty-two Susannah married Stan, a composer sixteen years her senior. Seduced by the vision of a bohemian future—he would write music, she would take arty photographs—she moved into his grungy loft in Brooklyn. To Stan's disgust, she got a job in the film department of the Museum of Modern Art, and loved it.

Susannah worked full-time, did all the shopping and cooking, managed the household, and handled child care for their daughter. Although her salary paid all his expenses (including child support payments from his first marriage—she even wrote the checks), Stan begrudged Susannah her professionalism and mocked her "bourgeois" values, although his principles didn't keep him from enjoying his in-laws' swank parties or Caribbean vacation home. At one point he sneeringly accused her of being "a different bag of goods than I bargained for. I thought you were going to be a photographer, and here you are in a career, actually liking it." Susannah's brow furrows. "It was a very weird thing, similar to a lot of his abusive stuff, because here I was supporting everybody—somebody had to—but I was made to feel horrible about it because I wasn't doing my art." It was also Susannah's fault that Stan wasn't composing, because only as a tortured, solitary soul could he be creative.

She and Stan never fought—"this wasn't a person you argued with," she recalls flatly—but one more put-down, over drinks at an outdoor café, led Susannah to pick up her glass and propose a toast to herself.

> I said, "I think I'm pretty terrific, I really do. Yes, I am making a career and a name for myself, and I think that's terrific. And I think it's nice that I'm able to support my family—and your two children—and I think you're full of shit." He didn't say a word. That night we were lying in bed and he curled up in this kind of fetal position and said, "I think this is going to lead to divorce." I remember at that moment, in the dark, it was like my hair literally stood on end, my eyes popped. *Divorce!?!* I had never thought of it, it had honestly never crossed my mind. Then, over the next fifteen minutes, I went from "*Divorce!*" to "Hmmm . . . divorce." It went from horror and disbelief to Solution! Joy! Rapture!

The next day she took off her wedding ring. This was twenty years ago, when divorce was more scandalous, but Susannah wasn't one to worry about flouting convention. She stayed put and tolerated Stan's abuse because she romanticized her much older mate's overbearing ways and because she had assumed responsibility for the marital maintenance contract.

Physical Abuse

Verbal abuse can pave the way for physical violence. Sometimes when abuse crosses that line—with a punch or a rape—it provides a clear signal for women that it's time to get out, as was the case for several women in this study. But even when abuse turns physical, some women can't leave. Criticism from husbands reinforces wives' reduced status and expectations, creating a vicious circle. Because their self-esteem is so low, battered women are often too ashamed to ask for help. Because many abusive husbands forbid their wives to work, these women often have no money. And because they fear violent retribution, they are often afraid to ask for help.

Violet was caught in just this sort of vicious circle. "When you have low self-esteem, and you don't feel good about yourself, the first person who comes along and says, 'Mmm, you look *nice*, I like your hair, that dress is really sharp on you,' you pick yourself up a little, say, 'Oh!'" says Violet. "That's what happened." Now fifty-three, at twenty-one she married William, the first man who made her feel that way. As a girl she had been taller than the other kids in her Peachtree City, Georgia, schoolyard, and mercilessly teased about her thick glasses and wandering eye. The glasses are still there and she is rail thin, but this soft-spoken woman points with pride to a photograph of three strapping boys who share her serious gaze.

Her fairy-tale fantasies went up in smoke right after their first baby was born, when it became obvious that her husband was no longer as attracted to her. "You feel bad, like what have I gotten myself into?" she asks rhetorically. The answer, as it turns out, was into eleven years with a man who cooked, shared the housework, and was a good father and decent provider, unless he was blowing his paycheck on a new girlfriend. She was crazy about him. He abused her terribly, verbally and physically. He told her she was stupid and ugly. "Why can't you be built like her?" he'd jeer, pointing out one of his more voluptuous girlfriends. When neighbors—"they could see me on the street with a black eye, or I'm limping down the sidewalk"—would comment, William told them, "She's crazy, she's putting on, she's losing her mind. You better stay away from her." Over the years Violet nearly came to believe it herself.

William started hitting her about five years into the marriage, when she first confronted him about staying out all night, and the beatings grew worse over the years. One night, finally deciding she had to witness his philandering with her own eyes, she followed him to a club and confronted him with a girlfriend. When William got home, he lit into her.

> He started hitting me, turned the mattress over on the bed, blood was everywhere. Then he knocked me down and he took his foot and literally stepped on me—like you have a cigarette butt and you're trying to out it? He just twisted his foot on my face. It was on the side where I had my bad eye, and then I had to go to the hospital. They gave me nothing to numb me, they just stitched it up, and the doctor looked at me and laughed, said, "What did your husband do, catch you out with your man?" My feelings were *so* hurt. It was horrible, it was horrible.

The extreme brutality of this beating, and the fact that two of her sons had witnessed it, were the catalysts Violet needed. "A week later my older son said, 'Mom, when I'm sixteen I'm going to get a gun and kill him,'" she recalls sadly. She went to the legal aid society in downtown Atlanta and filed papers. "I caught the 56 bus and cried the whole way home, because it was like my whole life was over except for my poor children, and no father with them. Look at the mess I made," she groans. "I just had a guilt trip running on me." Now Violet looks after children in a battered women's shelter, a job she loves. "I like the way I feel about myself when I go out that door," she says proudly, and she also likes being in a position to help women trapped in abusive relationships as she once was.

Like many victims of physical abuse, Sophia grew up witnessing her father drink and beat up her mother. In her own life, the pattern asserted itself on her wedding night. Her husband Mac's idea of foreplay was to slap his wife around. "It was like he couldn't get excited until he punched me around a little bit, beat me up, and beat me up he did." Black eyes, bruised mouth, cut lips, the works. The lesson of her divorce, Sophia says quietly, was that "no human being should be treated like an animal."

A small, stylish woman with copper-colored skin and shoe-button eyes, she pulled up for the interview in a gray Cadillac, wearing high heels and a tailored suit. "I would like you to use my real name, Sophia Gonzales Scherman—so they'll know the Hispanic part of me—so they know it happened to a real person," she instructs. Born in Wyoming to a Mexican family of migrant workers, Sophia and her family followed the crops around the

FACING DOMESTIC VIOLENCE

The place where women are most likely to experience violence is not in a dark alley but in their own homes, and not at the hands of strangers but at the hands of husbands or lovers. It's a high price to pay for intimacy. As many as one in seven women have been raped by their husbands, and one in four beaten at least once. Every fifteen seconds a woman is battered in the United States by her husband, boyfriend, or live-in partner, with four million incidents reported each year. Domestic violence does not discriminate, affecting women of all ages, races, and social classes.

"There are signs that show an abusive man," says Violet, who has learned them through her job in a battered women's shelter. Especially in the beginning, such relationships often appear very romantic because men are so attentive, but the objective is to isolate the woman emotionally and socially. "They don't want you around your family or friends, they want you all to themselves. If you don't do what they want, they holler, they scream, they threaten you. They abuse you, then you have a honeymoon period when they say, 'I'm so sorry, I'll never do it again,' send you flowers, candy, take you out. Then the minute they come home and the food's not hot enough or the shirt's not laid out that they wanted, they start all over again." In Violet's training, the pattern is called Power and Control, "because the man wants to be the powerful one. It's, 'Can we go out tonight?' 'No.' Or 'I need $10 to go to the store.' 'We're not going to the store.' He's in control, he's the powerful one."

Other common signs of a potentially violent man are extreme jealousy or possessiveness, control of money and decision-making, pressure to commit quickly to the relationship, cruelty to animals or children, and use of force during sex. If you are being abused, you can get help by looking in your local phone book under "Domestic Abuse," or by calling the National Domestic Violence Hotline at 800–799-SAFE.

prairie states and the Pacific Northwest, eventually settling in California when her father got a job laying track. They lived in a one-room, dirt-floored cabin, but, she says, "you know, I never knew that I was poor until I got older, when people told me."

Dissuaded from taking college prep courses because of her poverty, Sophia "did the normal thing: I got married right out of high school like the rest of the Hispanic kids do, became a wife and a mother." Mac was of Scottish descent but hung around with Hispanics and adopted their culture. Sophia was forbidden to work, to leave the house, to have a driver's license. Her husband told her she was too stupid, and that places wouldn't let her in because she was Mexican. He also told her she was ugly, illiterate, and no good. "That was part of the process of stripping me down so he could mold me the way he wanted me," explains Sophia, who endured for twelve years and through the birth of three children. "In my culture you do what the husband says, you do what the husband wants, and you are the obedient wife and have babies," she explains, "and that's what I did." Wife-beating was common, but "you wouldn't tell your mother or anybody. You just took it and that's the way it was."

It took three tries, all during the last three years of her marriage, for Sophia to extricate herself. The first two times she told Mac she wanted him out he didn't believe her. "I was just a little puny thing, it didn't come out real forceful," says Sophia, able to smile about it twenty-five years later, and adding, "but I was getting stronger each year." After his wife told him for the third time that the marriage was over, Mac climbed through the window and raped her before leaving for good. Sophia got a job in a dress shop and went to see an attorney when she found out she was pregnant. That day Sophia said to herself, "I don't care if he tells me it's going to cost a certain amount of money, that I'm going to be on my own, that I'm going to be responsible for these other people, I'm done. I'm out of here." She is infinitely glad she took the initiative and broke away from what she calls her "pretend world," when she would focus on the good times "because I was pretending that I had the perfect marriage. Outwardly everybody thought so." They never saw the black eyes because she never opened the door. When she finally got out of the relationship, Sophia felt "lucky to be alive."

Sophia and Violet had few material advantages, but it is not just lower-income women of color who experience physically abusive marriages. Even well-off white women can feel trapped in such relationships, Nicole Brown Simpson being a prime example. Although evidence of physical abuse silences the whispers of "How could she leave?" that many divorcing women overhear, an equally insensitive and accusatory question takes its place: "How could she have stayed in it for so long?" The burden is still the woman's. But there's a better question, which is not asked enough: why do men beat their wives?

EILEEN

I used to wish sometimes that I'd done it sooner, but I think that I've learned as I've gotten older that things take the time that they need. That you can't rush it.

The Rewards of Moving Out into the World

"I got my life back! I got my head back!"

The terror of deciding is real, and so is the risk: for a while, everything does go to hell. Whether or not they were abused, when women start putting their own interests alongside—or, God forbid, ahead of—their obligations to family and community, there goes the apple cart. When gender roles are shaken up, the whole social order *is* threatened. That's just fine, because it means that women stop being hostage to a set of destructive circumstances: insult, silence, and submission.

Lorraine, a second-grade teacher with a ready, crooked smile and a halo of frizzy blond hair, describes her married life as living with blinders on.

When I was growing up, when I was me, everything was possible. But these blinders had developed and that's the way I was living, knowing "There's stuff out there but I can't get to it." But as soon as I made my decision, it was like somebody took them off. And now, I see at an angle that begins at me and goes just *wide*.

Lorraine says jubilantly, "I got my life back! I got my head back!" She adds, "There are times when I look back and think,

'Why did I put up with this for so long?'" and recalls a bit of advice from a friend who has stayed in touch with both her and her ex-husband:

> Remember all that strength you used all those years to keep things calm in your house and make it work, even though it wasn't what you really wanted out of life? You don't have to use it that way anymore. Use that same strength now to get through the times when your kids are upset, or your ex-husband is acting strangely, or dealing with your lawyer. It's not more strength, it's strength you've had all along.

"I'm glad to have what's ahead to move on to," she says. "If I could give advice, I'd say, 'Be strong. You are strong.' There's a whole life out there."

One woman described the words of a friend unhappy in her marriage but at a loss as to what to do about it. "Sometimes I look out the picture window and I see the whole world passing me by," the friend said, weeping. "It makes me crazy." Each of the women in this book has found her way through that picture window and out into the world.

Rich or poor, black or white, mother or childless, once a woman no longer sees her life as a series of punishing choices and crippling contradictions, she is free to inhabit it fully. She needn't stop caring for those she loves; she can start taking care of herself as well, in ways more of her choosing. She can stop worrying about being good enough to be loved. She can stop holding back and move powerfully ahead.

Laws and Lawyers

W hen the tapestry of a shared life is unraveled, so are the identities and illusions that maintained it. Even when both parties are in accord, the upheaval is traumatic, cumbersome, and costly. The entry into unknown territory with such high stakes is frightening, the language of the law and hostile posturing of attorneys intimidating. Handing over a once-intimate relationship to an impersonal legal machine feels alien, and the prospect of having the most private affairs become a matter of public record is repugnant.

To a considerable extent these apprehensions are justified, even for women who can afford a really good lawyer. But there are many ways to work within the system to get a good agreement, and many original solutions too. Courage, perseverance, and deep pockets all help.

The Shortcomings of the Legal Process

IT DOESN'T ADDRESS THE EMOTIONAL ASPECTS OF DIVORCE

The legal process does deliver the goods—it dismantles the marriage—but it's little help with anything but the bare mechanics.

Lawyers are trained to be their clients' advocates, to look at the facts from one point of view. They're not trained to cope with men and women in crisis. To them, divorce is a business deal. It's a lot about money and a little bit about fairness, and not at all about anger or fear or grief or revenge. It dissolves the union, but it doesn't begin to address the welter of emotions stirred up by signing out. Separating emotionally is as important as separating legally, but those who try to use the law to effect their emotional divorce are doomed to dissatisfaction.

Few lawyers are aware of or willing to deal with the feelings the paperwork doesn't address. Yolanda's lawyer at least acknowledged them. "She said that she knew how important the emotions were in my life, but that they had no place in my thinking about this"— easy advice to give and hard to follow, notes Yolanda, the public relations director. I got much the same advice, in the form of frequent admonishments from my attorney to calm down and sit tight, but found it impossible to obey.

The emotions matter—and matter deeply—but it is up to the divorcing woman to decide how much weight they should carry when it comes to strategy and decision-making. Whether out of guilt, feelings of independence, or the desire to pacify their mates, initiators are likely to ask for less. "I didn't want anything from Sol," explains Esther, "because I was the one who left him." Recognizing their lawyers' bias or tunnel vision, many women at times ignored their counsel in this regard and were not sorry. Several, though they earned much less than their husbands and were not legally obligated to split the children's college expenses, opted to do so out of pride. It was financially difficult but emotionally necessary.

"I just felt so independent, and by golly, I hated the idea of alimony and being tied to him," says Wendy. Wendy's income as a cookbook writer and columnist was a fraction of what her cardiologist husband earned, but it was enough to live on. She and her husband sold their big house and split the proceeds, and against the advice of her attorney Wendy signed away her rights to the rest of their property and to his pension fund. "She was a good attorney; I was a bad client," acknowledges Wendy frankly. The strategy worked to her advantage a few years later, however—as even her attorney would admit—when Jim ran into some problems with the

IRS and wanted Wendy to get involved. "Repeat after me," said the lawyer: "You tell him, 'I didn't mortgage your future, you take care of the past.'"

THE BUSINESS OF LAW IS ENTIRELY SELF-REGULATED—AND FULL OF CONFLICTS OF INTEREST

Unlike, say, used-car salesmen, divorce lawyers aren't subject to the kinds of safeguards that protect the consumers of other types of services. In most states lawyers aren't legally required to tell their clients how much they think a case will cost or how long it will take. One reason is that ignorant or intimidated clients make their lives easier. Another, more legitimate reason is that unlike a used car, which has a book value, a divorce has too many variables to have a set price and duration. The only real rule of thumb is that litigation is lengthy and expensive.

The law remains very much a male-dominated field, although much less so in family law than in other branches. An old-boy network forms a closed system, in which important business is often conducted privately. Any reviews of standards or conduct are carried out by fellow lawyers and judges, who may have a considerable stake in the outcome.

The longer and more complicated a case, the more a lawyer can charge, so an unscrupulous attorney will often drag out proceedings. Lawyers don't have to answer for their own mistakes unless their clients sue them for malpractice. Another potential conflict of interest occurs when the spouse designated the "moneyed" party is ordered to pay the legal fees of the "needy" spouse, usually the wife; since the husband is paying the bills, the wife's lawyer may have a vested interest in pleasing him. However, such court orders, designed to level the playing field, give many wives a crack at a fair settlement that would otherwise be beyond their means. But no matter who wins or loses, the lawyer gets paid.

JUDGMENTS CAN BE UNFAIR OR HARD TO ENFORCE

Ten percent of the women in this survey were awarded alimony or maintenance, compared to 15 percent of divorced women nationwide, and for two of the five, payments were for a fixed term of five years or less. So much for the stereotype of the greedy woman bleeding her ex-husband dry. Some forfeited alimony because they wanted out at any cost, and many because they simply didn't think their ex-husbands would ever agree to pay it. A number didn't want the money, some because they didn't want the ongoing financial dealings, others because they felt they were not entitled to it.

Thirty-eight percent of the women in this survey are among the 48 percent of married women nationwide who bring in half or more of their household income. Francine, who married a Danish graduate student, was among them. "I was scared shitless he was going to ask *me* for alimony," she admits, eyes widening at the memory. (No husbands in this study did get alimony, though in several cases it was a distinct possibility.)

For the great majority of women with young children, child support is inadequate or not paid at all. It never occurred to these women not to provide for their children, and the fact that this was not true of their children's fathers angered and surprised them. Most lacked both the money and the energy to seek legal remedy to get the child support they were owed, and felt their ex-husbands would continue to evade their responsibilities in any case. Delores hasn't gone after the $250 a month Carlos owes her in child support, "because what do I get? To spend more money and get more headaches?"

Experience bears out this sorry conclusion. Only one woman interviewed pressed the matter in court, and after three years she too gave up. Some, like Ginny and Karen, settled for a nominal amount of child support that they felt they had a reasonable chance of actually collecting, instead of pressing for 50 percent of the actual expenses of raising a child. These women and children, who make up the great majority of America's growing percentage of single-parent homes, are those whom the system most conspicuously fails. The needs of the children are often the first things to go by the wayside in a battle between husband and wife.

DIVORCE PROCEEDINGS ARE INHERENTLY ADVERSARIAL

This regrettable fact often has disastrous consequences for both spouses, not to mention any offspring caught in the crossfire. For the lawyers, the emphasis is not on fairness or compromise, but on getting the best possible deal for one client and discrediting the other. In general women don't like the rules of combat, they don't want to learn them, and they aren't sure whether they have to abide by them.

Lorraine opted not to. Her lawyer was an old-fashioned type who pointed out that Jim, who never even hired counsel, was in default. "Throw the book at him. File for everything you want," he urged. "At that point I was ready to punch the lawyer, or cry," Lorraine recalls exasperatedly.

> I remember standing up and saying, "Look: no, this is what I'm doing. You just need to tell me if any of us is going to be raked over the coals by what we're trying to do. Have I provided for the kids properly? Is there anything he can come back at me for? Is there a better way that we all can do this?" He finally got the gist of it.

She ended up with what she feels is a fair settlement.

Like Lorraine, almost all of the women interviewed put a high priority on keeping things amicable during and after the proceedings, with varying degrees of success and cooperation. Many women do so for the sake of the children. Some simply can't tolerate hostility. Women, who are used to viewing themselves in relation to others, often feel that when a relationship ends, a piece of themselves goes along with it. For these women, maintaining some connection with their husbands is the only way they can sever marriage ties. They may ignore pressure from their attorneys to act more aggressively because the material gain does not justify the emotional loss.

Finding a Lawyer

Because almost every issue is emotionally charged, few couples can actually sit down and work things out without the help of a skilled intermediary—especially if they couldn't resolve major issues while they were still partners. Good representation is especially important given the fact that contemporary divorce law leaves broad discretion to lawyers and judges in applying and enforcing child support and asset distribution guidelines. A lawyer who is too belligerent or too passive can do more harm than good, but a competent lawyer can be a godsend. In particular women with overbearing husbands or women who are financially naive benefit from having a trained advocate. Phyllis, a fifty-year-old court stenographer, and her two small children moved into her mother's apartment when her eight-year marriage broke up. When her not-yet-ex-husband would come over and browbeat her, "I learned to be a broken record," Phyllis recalls. "My lawyer taught me to say, 'I have to talk to my lawyer,' 'I have to talk to my lawyer,' 'I have to talk to my lawyer.'"

Women with limited resources are less attractive to top-flight lawyers, who routinely require an initial retainer of $3,000 to $10,000. If the settlement looks to be worth their while, some are willing to be paid at the end of the proceedings. Matrimonial attorneys have a bad reputation, but most complaints I heard were about garden-variety ineptitude. Some lawyers do patronize or take advantage of their clients, but others dispense moral support and timely counsel—for a price. Fees range from $125 to $400 an hour, and only the Kleenex is free.

Finding a lawyer turned out to be a haphazard business for the women I interviewed. One way or another, they found representation, relying on everything from word of mouth to television advertisements. Quite a few deliberately sought out female lawyers. Unsurprisingly, gender was no guarantee of brains or principles, but many felt more comfortable with a woman advocate.

The client-lawyer relationship is match of personality and style like any other, but the client is extremely vulnerable. In fact, if her attorney is male, she may have romantic fantasies or erotic dreams about him; he, after all, is the hero squiring her through chaos to security—a white knight with the meter running. It's also possible to find the right lawyer at the wrong time. Some women are unprepared for an attorney's combative posture at the outset of the proceedings, though they may look back and wish they had followed that initial advice. Even the right lawyer can be rendered helpless by intransigence or vindictiveness on the part of either spouse. And some attorneys don't get the job done.

Yolanda fired her incompetent first lawyer, and fared much better with a second who knew the local judges well. Nancy went through three lawyers, firing the first when she found out the woman was consulting with her husband ("Well, he won't hire anyone," was the attorney's explanation)—and billing *Nancy* for the time. After a long conversation in which the second lawyer basically admitted that she didn't know what the hell she was doing, Nancy dismissed her. "She socked me with this kind of subpoena for money and I immediately wrote her back and said, "You, if anybody, know I don't have the cash. If you want to arbitrate, fine," recounts Nancy. "I never heard from her again." The third lawyer came in and expertly wrapped things up, but Nancy's faith in the legal system is considerably the worse for wear.

Interviewing more than one attorney paid off for me and for other women I spoke to. I first consulted a slick midtown attorney, but his reputation as a ruthless litigator unnerved me; I didn't want to go for blood, or put my husband unduly on the defensive. I went looking in a thoroughly scientific manner for someone else: a mother in the schoolyard mentioned that her downstairs neighbor was a divorce lawyer and I called him up that night. Within min-

---

IF YOU DON'T KNOW HOW TO FIND A LAWYER

Ask divorced friends or neighbors who represented them, and whether they were satisfied. Your local bar association may offer referrals, as will many women's organizations. Most communities are served by one or more legal services program offices that provide free legal services to people with little or no income. Consult the community services pages of your local phone book or look in the white pages under "Legal Services of . . ." or "Legal Aid." There's always the Yellow Pages, under "Lawyer." If you live near a university, find out which law professors are interested in women's rights and see whom they recommend. You can check a lawyer's track record by looking up case files in the courthouse and calling litigants who have used his or her services. Two Internet sources that provide nationwide directories of attorneys and legal specialties are LawyerSource (http://www.lawyersource.com/) and Attorney Finder (http://www.attyfind.com/). On America Online, type keyword "Legal," or, ironically, "Wedding," to reach free legal advice on divorce and custody issues.

TO PROTECT YOURSELF AND GET YOUR MONEY'S WORTH

- Don't sign or agree to anything you don't understand. Your lawyer should explain what's happening at any point in the proceedings, and should write any documents in plain English.

- Learn what you are entitled to. Your lawyer is not responsible for educating you, so educate yourself. Your lawyer does not have the right to demand a nonrefundable retainer at the beginning of the case or a bonus at its conclusion; to withdraw from the case without just cause approved by the judge; or to force you to settle. (A complete Bill of Rights for Divorce Clients, written by Karen Winner for the New York City Department of Consumer Affairs, is available in her book *Divorced from Justice: The Abuse of Women by Divorce Lawyers and Judges* (New York: ReganBooks, 1996, pp. 208-10).

- Interview more than one attorney. Don't hesitate to bring an objective friend along. Ask if the lawyer has handled other cases like yours, and whether he or she has litigation experience. Get an estimate of how much it will cost and how long it will take, and agree on how the attorney is to be paid. Obtain the attorney's assurance that you'll be informed as soon as any retainer is used up. Find out whether junior associates will be handling aspects of the case, and what they cost. Then get it in writing; a retainer agreement is legally binding.

- If a lawyer is vague or evasive, find a different one.

- Get copies of correspondence and relevant documents.

- Request regular, written, itemized bills, and go over them carefully. Get an explanation of any unclear figures or terms. As best you can, make sure any research, legwork, faxes, and phone calls charged to you relate to your case.

utes I was in tears, because I was such a mess and he sounded so nice. He *was* nice, for which I was grateful. He put a high priority on the children's interests, and because I never had to argue with him about what the moral course of action would be, my conscience was spared a lot of wear and tear. He dispensed some excellent advice, including an early tip to keep custody issues separate from financial ones.

The Mechanics of My Divorce

Unhappy about the possibility of spending thousands of dollars on legal fees, my husband wanted do without lawyers. I was game, but told him I wouldn't negotiate directly with him if he insisted on moving back into our apartment. (He had moved out on a trial separation basis, and when he announced six weeks later that he was moving back, I told him that I had decided I wanted a divorce.) He came back, on the advice of a lawyer as it turned out, probably because his having an alternative residence would have made it easier for me to establish that I was the primary residential parent. After four interminable weeks, during which I refused to deal and he refused to retain counsel, I raided the bank account and served him with papers in front of our apartment. It was the pits, but it got things moving.

New York is not a no-fault state, so I was required to choose grounds for the divorce. "Cruel and inhuman treatment" was the least Dickensian. I then had to substantiate the charge by dredging my memory for every remotely unkind or irresponsible act my husband had committed over the past decade. The portrait that emerged was not inhuman but all too human. He was no monster, just a jerk; I was no saint, just a wife who helped him be one. (Reading the papers must have been horrid for my husband. When Hillary, a woman I interviewed, read all the inaccurate accusations her husband's lawyer had coached him to come up with for their proceedings, she recalled, "I felt like I had been raped.") The process was specious, stupid, mutually debasing, and utterly demoralizing, and when I sobbed to my attorney that I felt perfectly vile, he said calmly, "That's what everyone says at this point." New York is also an equitable distribution state, which meant that, among other things, because an apartment purchased solely with my earnings was held in joint title, my husband received half its value.

My legal expenses were significantly higher than my husband's, because I was desperate to move things along and hopeless at following my lawyer's advice to chill out. He drafted all the papers and made most of the phone calls, and my husband and his attorney could sit back and wait. If I had held out, I might have persuaded a judge that more than 50 percent of our assets should be mine, but I would probably have spent the difference in court costs, damaged my children by increasing the duration and hostility of the proceedings, and permanently embittered my ex-husband in the process. The fact that I'm now in debt pisses me off every so often, but I'd sign the papers again tomorrow.

As divorces go, mine was a good one. My ex-husband and I have a lot of dealings because of our children, and try to oblige each other for their sake and our own. But even a "simple" divorce is complex, and even a "good" divorce is painful. All it takes is one spouse who obstructs the process for it to drag on for years. All it takes is one unscrupulous lawyer or a spell in court for it to devastate financially. And no matter how binding an agreement, it's up to the two people who signed it to abide by it, and largely unenforceable otherwise. That doesn't keep people out of the courts; Americans have always been inclined to sue for what they believe is rightfully theirs, including the pursuit of happiness.

How American Divorce Law Has Evolved

Following the English model, the adversarial nature of divorce was established early on in America. In colonial Massachusetts penalties for violating the marriage vows included whipping, fines, being locked in the stocks, and more often than not a prohibition on remarrying (so the maritally impaired would not go on to create another damaged family unit—family values, Puritan style). Not that matrimony afforded women much protection: once married, a woman ceased to exist as a separate legal entity, yielding ownership of her body and personal property, and the custody of any children to her husband. Well into the 1800s, when she married a woman exchanged autonomy for support. Good bargain or not, the arrangement ensured her financial dependence.

Unhappy couples had very limited legal recourse until after the Revolutionary War, when grounds for divorce were formalized. As the century progressed and records began being kept, the number of divorces grew steadily, increasing from almost 10,000 divorces in 1867 to 72,000 in 1906. The idea that the marriage vow should be a dissoluble contract gained in popularity. In 1868 women's suffrage leader Elizabeth Cady Stanton declared in *The Revolution*, a women's rights journal that she edited, that each "slave" who fled "a discordant marriage" was a source of joy to her, and predicted that "with the education and elevation of women we shall have a mighty sundering of the unholy ties that hold men and women together who loathe and despise each other."

In the last decades of the nineteenth century, fueled by the suffrage movement, married women were granted the right to own separate property. Laws were liberalized to recognize a broader range of grounds for divorce, and custody laws changed too. By the end of the nineteenth century, as men moved into the work force and left women responsible for child care, the presumption that young children were best off staying with their mothers replaced the father's absolute right to custody. Concerned about the divorce "epidemic"—rates were particularly high in the American West, where divorce laws tended to be more liberal and residency rules less strict—conservatives fixed their sights on restricting those very laws. State legislatures passed more than one hundred restrictive laws between 1889 and 1906, with South Carolina banning divorce outright.

Overall, though, conservatives' attempts to restrict divorce met with little success. Divorce accelerated from the second half of the nineteenth century onward, and as ever most petitioners were female; more than two-thirds of all divorces between the 1880s and World War I were granted to women. Booms marked both postwar periods, and each time the rate stabilized at much higher levels. This was generally attributed to the growing economic emancipation of women, just as the jump in the early 1970s—doubling between 1966 and 1976 alone—was blamed on the women's movement. Rising expectations of marriage and a decrease in the stigma attached to divorce were certainly factors. Many divorcing women were mothers, a significant trend that increased throughout the 1980s. Not until 1970, though, was divorce legal in all fifty states.

The Implications of No-Fault and Community Property Legislation

In the 1970s lawmakers came up with a radical plan to temper the adversarial nature of the process by permitting divorce by mutual consent, without one spouse being designated the injured party. By 1985 most states had adopted some form of this "no-fault" model. (Half still kept the traditional "fault" system on their books, with no-fault only as an option.) Because it was no longer necessary to assign responsibility for the dissolution of the marriage, one spouse no longer had to sue the other, which made the whole process more humane, simpler, and much less expensive.

Along with no-fault legislation came equitable distribution and community property laws. In the 1970s nine states, with California paving the way, passed community property provisions intended to allocate marital resources more fairly. (Community property laws assume joint ownership of all assets acquired during the marriage, and call for a fifty-fifty split. With equitable distribution, the apportioning of assets is proportional to the contributions of each spouse, including the "non-economic" contributions of mothers and housewives.) Olivia, who spent her married life working inside the home, still sounds vaguely disbelieving as she quotes the message her lawyer drilled home.

> She said, "This money that you're asking for is yours. You don't have to ask for it, it is yours by law, half of everything that's come into the marriage in thirty-five years." That was the biggest thing. I kept saying, "I'm not going to ask for this," and she'd say, "Forget ask. It's yours by law." That made me feel like a million bucks, because I felt like I had earned every bit of it. I had raised five kids, essentially by myself.

The idea behind these laws was to acknowledge the nature of marriage as a financial partnership as well as a social one. The legislation was intentionally designed to protect women and children financially, and inequities are no longer built into the laws themselves. But because lawyers and judges whose attitudes may not have evolved have great discretion in applying the laws, the protection they afford is uneven.

Another brief upswing in the divorce rate immediately followed the 1970s legislation, but most historians attribute it to the speeding up of cases already in the pipeline. "Most evidence suggests that no-fault had little effect on divorce rates [which] were going up anyhow," says sociologist Andrew Cherlin of Johns Hopkins, adding that "the great misconception is that divorce laws change people's behavior. People's behavior changes divorce laws." Divorce historian Roderick Phillips concurs: "If anything, the divorce law reforms were a response to the rising incidence of divorce rather than the cause of it." After all, the surge of divorces in the 1960s well preceded no-fault legislation.

The American divorce rate has in fact declined slightly in recent years, as has the number of new families forming, a trend that worries many social scientists. Nixing no-fault is hardly likely to encourage the marriage-shy, as Hanna Rosin suggests in the *New Republic,* a point that the matrimonial proponents might do well to consider.

Back to Blame?

Led by social conservatives, a backlash against no-fault divorce is now in full swing. Their complaints about how divorce has gotten "too easy" have a familiar ring to historians who recall a similar outcry in the late nineteenth century, when Horace Greeley, for example, complained that too many people were getting "unmarried at pleasure." Measures currently before a number of state legislatures would revoke the no-fault option in cases where one spouse opposes the divorce. A front-page article in the *International Herald Tribune* from the *New York Times,* aptly subtitled "Blame Is Back," describes it as "an emerging campaign to restore notions of guilt to divorce law." It's based on the notion that in a divorce, somebody has to be the bad guy. This is wrong and wrongheaded, but the notion of retribution has endless appeal to the self-righteous.

Another punitive fallacy is that the typical divorcing couple consists of self-indulgent Me-Generationers who wake up, decide they can't stand his snoring or her bitching one more day, and casually call a lawyer. "It's so intuitively obvious," says David Blankenhorn, president of the Institute for American Values. "If you make it quicker and easier to break a marriage commitment, especially when either party can do it at will, then it's logical that more people will take that opportunity." Obvious it may seem, but it is not so. Divorce is not easy, or popular.

The women interviewed in this book agonized long and hard over the decision to end their marriages. Even those in the meanest of marriages wept at the loss to their families and the shattered dreams of intimacy and happiness. In an essay on her divorce, Ann Patchett proposes that the divorce police be redeployed:

> Make marriage harder if you want to. Outlaw those Vegas chapels with the neon wedding bells, require marriage applications modeled after tax forms, but leave divorce alone. It's grueling. I have never known anyone who went into a marriage thinking that they would have to get out, and I have never known anyone who got out simply. To leave you have to involve the courts. You have to sue the person you live with for your freedom. You have to disconnect your life from another life and face the sea alone. Never easy, never blithe. Never.

Divorce is punishing enough. The stigma of being the "bad guy," or the one foolish enough to have tolerated spousal abuse, scars needlessly.

The third fallacy about ending no-fault is that upping the legal and logistical ante will somehow make unhappy marriages more endurable. Durable, perhaps; the threat of an ugly, protracted legal battle does indeed paralyze some spouses, who are afraid to move on no matter how miserable their domestic lives. What ending no-fault would guarantee, however, is an increase in desertion, fraud, and the tragic hostilities that can turn divorce proceedings into scorched-earth campaigns. Assets are spent on lawyers instead of building new lives or providing for children, a real irony given the pro-fault movement's "pro-family" stance. Energy goes into excruciating struggles over children and property, instead of into figuring out how to maintain decent relations with the person who once mattered most in the universe, and to moving on. Children suffer prolonged exposure to the most damaging possible circumstances—parents who fight—and their deliverance is left in the hands of strangers and overburdened courts. Fractured into warring camps, families may never recover.

Ironically, some women's rights advocates also support the repeal of no-fault. They point out that equitable distribution settlements often do not split along gender lines as originally intended, my divorce being a perfect example. The old laws linked property to "fault," forcing the divorcing wage-earner to continue to support his family and giving "innocent" wives considerable leverage in negotiating settlements. The loss of this bargaining power concerns women's rights advocates, but their automatic assumption that wives are victims does women no favors, and is unfair to the many "innocent" husbands out there whose wives leave them. The underlying notion of innocence versus guilt should be jettisoned. It reinforces the age-old link between goodness ("innocence") and passivity, a big step backwards for authentic women's rights. It also completely disregards the fact that divorce is twice as likely to be initiated by the wife as the husband. These ways of thinking are truly sexist. When it comes to alimony and child support, income, not gender, is what matters. Laws that punish the initiator should be abolished; they hurt women as much as men.

Then there's ethics. The thought that someone would want to stay married against his or her partner's desires runs contrary to any humane notion of how people who once cared deeply about each other—and may still—should treat each other. It also runs roughshod over the cherished American right to self-determination and, for better or worse, exposes the happily-ever-after ideal as romantic claptrap. Inertia is indeed the ruling principle of many marriages, but if one spouse wants out it is no longer a marriage in the true sense of the word, no matter who makes the first move.

The Effects of Being the Initiator

SUSANNAH

I think that it's certainly easier from a how-you-feel-about-yourself standpoint to be the leaver.

In our culture, men are supposed to initiate and women to respond. That's how traditional relationships work, from first flirtation onward: the husband protects the "good," dependent wife.

It's bad girls who take the initiative (and not just in bed), so when a wife makes the first move, she forfeits her goodness and the protection it affords. Angry or threatened, many husbands lash out or dig in their heels.

Keiko's husband told her, "You betrayed me. You *left* me. You *abandoned* me." Keiko was better than Megan at shrugging off the guilt trip, but Rick's bitterness did impede her efforts to keep things amicable for the sake of their two-year-old son. "It didn't work," she says frankly. "He was angry all the time about being rejected." But Keiko was angry too. A striking, fine-boned woman of Japanese-American descent, she had married at age twenty-four, and divorced eight years later. Rick worshipped her, and his stability and sizable income as a lawyer and lobbyist for the Washington State lumber industry were appealing, but after the wedding he worked longer hours and grew more possessive and dismissive. Keiko felt increasingly discouraged. "I wasn't as important," she maintains. "I never was, during the whole course of the marriage. When I was pregnant I thought, 'This is ridiculous. I'm not valued, and I deserve better than this.'"

"You're not supposed to leave the marriage," Keiko observes. In her experience, "the man can beat you up and beat you up and beat you up and if he leaves the marriage you come out victorious, but if you take the initiative you're seen as being somehow wayward." Because she was not the victim of obvious abuse, she thinks her lawyer perceived her as less needy, even though her husband made more than twenty times what she earned as a fledgling real-estate agent. In her plush Seattle suburb Keiko has observed plenty of marriages coming apart, but very few wives choosing to move out of their well-feathered nests. Keiko managed to hold on to the house, and to establish herself professionally in a new field. For a while Rick dragged his feet, but as he saw Keiko establish her emotional and financial independence he eventually stopped fighting the inevitable.

SHARON
Being the one to leave gave me a sense of control of the destiny of the relationship. I felt I could continue floating through this sad cloud of nothingness, or I could take the bull by the horns and get on with things.

The wife who leaves takes control of the relationship, even if only to end it. This turning of the tables is particularly difficult to handle for those men who feel a need to be in charge. Yolanda, the chain-smoking PR director, was married to a civil engineer who systematically imposed his black-and-white views on Yolanda, keeping her as dependent on him as possible. When she told Steve she wanted a divorce, he was "furious, absolutely *furious*. He was as punitive as he could be. He was evil in intent and in speech." With a shrug Yolanda puts her finger on the sorry dynamic common to many divorces: "He liked being the big man taking care of the little woman, but once I wasn't the little woman anymore, he just really wanted to punish me."

Savvy and determined, Yolanda went to a lawyer before she left so she wouldn't do anything stupid, "and then proceeded to do everything stupid because I needed to get out of there." She moved because she couldn't afford to keep up the nice house in Princeton unless there was some arrangement for child support, "and he wasn't having any of *that* bullshit—though we did work out a custodial arrangement fairly quickly." Yolanda paid $3,000 to a lawyer whom she fired when he couldn't get child support for her eleven-year-old, severely learning-disabled son, Will. "I said, 'There's something wrong with this picture,'" she recalls grimly.

Less interested in money than in a solid arrangement that would hold up over the long term, Yolanda paid $4,500 to a second lawyer, who was able to work out a decent agreement in less than a year. She and Steve no longer argue about money because Yolanda has disengaged. "I don't play," she explains. She also feels that her initial concessions paid off in the long run. "I actually have a slight advantage, in the sense that he knows that he got away with not murder but something close, on the agreement. So when something special like camp comes up I generally say, 'Well, I think you ought to do 60 or 70 percent of that,' and he often says yes."

Being the initiator was a strategic disadvantage for these three women. From a negotiating point of view it's always a disadvantage to want something more than the person on the other side of the table does; it certainly cost me plenty. It can also take a psychic toll by inducing guilt and inviting blame. But from a psychological point of view being the initiator was an immense advantage for the majority of the women I interviewed. Scary though it was, taking charge of their destinies was empowering and exhilarating.

LORRAINE

Initiating made me stronger. I started this and I was going to see it through. I also got a lot of support from other people, who said, "Good for you!"

GLORIA

I liked being the initiator. I didn't want to wait around for someone to serve papers to me or anything. And I had a head start. I took $5,000 out of our mutual savings account before anything happened, so I could pad my pocket a bit for times when I didn't know what I'd be having.

Initiating has practical advantages as well as psychological ones. Going first puts a woman in a position of strength, because she can research her options at length and at her own speed. Yet many women who end their marriages are loath to exploit this advantage. Despite the possibility that mention of the D-word will put husbands into a defensive mode, many women focus instead on protecting their husbands' feelings. One of the first steps in the legal process is "discovery," in which marital assets are determined and totaled. Faced with the prospect of divorce, husbands tend to get to work immediately calculating their assets and protecting their interests before negotiations begin. Women should not be reluctant to take the same step; they too need to come to the negotiating table fully informed about the family's financial situation. Women should do their homework before making any major moves, regarding custody issues in particular. "I don't mean be sneaky, but be smart," advises one mother, "because things can be used against you and you've got to keep your cool at this point."

Deal-Making: Different Ways to Work It Out

Every deal is different. Several divorces in this book were simple, do-it-yourself affairs, short agreements signed on the kitchen table.

Very few women ended up in court or in protracted legal battles. The median duration of their legal proceedings was one year, the median cost $1,000, with Amanda's $60,000 divorce at one end of the spectrum and seven women at the other end paying nothing at all. If a business or a significant amount of property has to be allocated it gets more complicated, and the presence of children ups the ante tremendously.

The nature and duration of the negotiation depend on a number of factors: how much stuff has to be divided up, whether custody and child support come into play, and the personalities involved (including the lawyers'). A divorcing woman's best strategy is to take a hard look at the man she's dealing with. She needs to ditch her illusions and denials (easier said than done); to be realistic about his strengths and shortcomings; and to use her instincts and intimate knowledge of his nature to gauge his reactions and do what's right for the long run.

Theresa, the green-eyed nurse from Florida, had a bullying husband who was not just an attorney but a matrimonial attorney. When he saw that Theresa's decision was final, he figured he'd handle their divorce and the arrangements for their three daughters. When he handed her the agreement he'd drawn up, ordering, "You'll look at it, you'll sign it," she knew it meant trouble. She said, "Look, David, I don't want to have to be *nice* to you because I need something. I have rights. I have begged you for money for the last ten years, and I'm not going to do that anymore. I'm going to get a lawyer." Unfortunately, getting separate counsel didn't resolve much. David hated the woman Theresa hired; both lawyers dug in their heels. Nine months and $11,000 later, with no agreement in sight, Theresa realized nobody was operating in the best interests of her or her three children. When she announced her intention to go to court, David, to Theresa's surprise, called her brother. She laughs at the memory. "When David went to see him, my brother said, 'She's right. I got to tell you, man to man, you can't take care of a *fish*, let alone three children.'" Three hours later they reached an agreement. Theresa's brother, who runs a pizzeria, got her a more generous settlement than any other woman in this book.

Theresa was able to face her husband down and take a firm stand, but many women have a hard time facing the fact that anger and retribution will almost certainly have their day when marriages end. In retrospect Amanda has only one regret: "that I allowed myself to be terrorized by conflict during the marriage." The same fear cost her dearly during her divorce, both financially and emotionally, and she was not alone. Socialized to be nonconfrontational, many women accommodate or back off, even when it jeopardizes their best interests. They seize on acting conciliatory as a way to preserve the illusion that progress is possible without causing pain. It's not much of a strategy, but it often makes possible a first step in getting the legal process off the ground. It's a way to act without forsaking the moral high ground, and it maintains the traditional balance of power a bit longer.

Sharing a Lawyer

One way around the inherently antagonistic move of hiring a lawyer is for husband and wife to share one. Because it makes her seem less like the initiator, it's one way for the divorcing wife to hedge that particular bet. It also saves money. But there are several good reasons that hiring separate counsel is always advised, the underlying one being that a single person cannot fairly represent two sometimes competing sets of interests. Other reasons, such as the fact that a lawyer can only appear on behalf of one client, are technical. Sharing counsel also opens the door to future renegotiations because one party may subsequently claim to have gotten a raw deal at the hands of a lawyer who was not impartial.

Mediation

Mediation calls for a neutral third party to meet with the husband and wife to work out an agreement in a cooperative manner. It's an efficient and compassionate way to bypass the adversarial system, and when it works it's extremely cost-effective. Very few of the women in this book tried it, though. Either it didn't occur to them, or they figured they'd get the short end of the stick, and their concerns are not without justification.

Mediation occurs in private. If browbeaten during her marriage, a woman may be bullied during negotiations; if financially disadvantaged, she may feel pressured to settle early; if ill-informed, she may not understand what she is signing. (Of course these are handicaps in any kind of divorce negotiation.) By definition, a mediator must remain completely neutral during the negotiations, and is not required to educate the parties about their rights. The profession is unregulated and requires no knowledge of the law on the part of its practitioners.

Mediation is not legally binding, but a mediated agreement can become so when signed by both parties and incorporated into the judgment of divorce. Some states require mediation in custody and visitation disputes, which makes a lot of sense because the mediator can keep the discussion focused on the child's interests. Also, some lawyers and family therapists offer mediation as part of their practice.

Mediation worked for Hal and Tara, a massage therapist from Boulder, who went to a group specializing in family counseling and divorce mediation. Their agreement cost approximately $800 and took only six months or so to write, in part because there was no property to speak of. "They looked at the whole thing, and it required that both of us participate in this process all along the way," Tara explains. "It should be standard, because it was wonderful. And every time a confrontation came up, they were really able to diffuse it and bring the focus back on the fact that what we were really talking about here was our kids."

For Virginia, a cartographer who looks uncannily like Sissy Spacek, marriage counseling led to mediation, but "everything Chris agreed to in marriage counseling would last about a week, and then the next week he'd break it," including pressing himself on her sexually. At that point she gave up and served papers. Another woman had the same problem when her husband wouldn't sign the deal worked out in mediation. For them mediation proved a needless expense and a fruitless intermediate step.

Mediation is great when it works, but it requires three cool and mutually committed heads, and few couples on the brink of civil war can summon up the necessary detachment or collaborative spirit. For some it's a last-ditch recourse when attorneys are deadlocked on a particular point. If it fails, the couple may have to renegotiate the same points in court, a costly alternative, and a judge's perception of a spouse as "uncooperative" may work to his or her disadvantage in the settlement.

For my part, all I could think about was moving on as fast as

INTERVIEW A MEDIATOR AS YOU WOULD A LAWYER OR COUNSELOR

Ask why the person became a mediator and how he or she envisions working with the two of you. How much experience does the mediator have in this field? How much does the mediator know about family law, and what role will legal precedent play in the process? Will the mediator encourage the two of you to communicate outside the mediation process? Will the mediator want to meet with each of you separately, and if so, why? These questions don't have right or wrong answers; what's important is that you feel comfortable with the responses and the philosophy behind them, and feel that the mediator's method will suit you and your husband. Trust and rapport are crucial. Ask, too, how the mediator feels about your using consultants or lawyers. (A mediator should be open to clients being well informed, but it is not the mediator's job to meet with your lawyers, or to hold meetings at which you are not present.)

The Academy of Family Mediators, based in Lexington, Massachusetts, offers a nationwide referral service of practitioner-members who meet its professional standards. The number is 617-674-2663.

possible to the next stage of my life; mediation didn't even occur to me. I also felt I needed an advocate, someone who would keep my long-term interests in mind in a way that I was not capable of doing. Accompanied by our respective lawyers, my husband and I did have several productive four-way meetings. It wasn't exactly mediation, but it was an effective compromise.

The Pros of Being Conciliatory

From both a moral and practical point of view, it's obviously better to keep divorce negotiations as friendly and low-key as possible. That was a top priority for the majority of women interviewed, and for many it paid off emotionally and financially.

Winifred, an artist and licensed therapist from Baltimore, never considered mediation but was nevertheless intent on staying friends with her ex-husband and stepdaughters. That possibility made the divorce psychologically bearable, in part because, by her own admission, she was "terrified of anger." Forty-five, with spiky bleached hair and stylish horn-rimmed glasses, she was thirty-five when she divorced a very wealthy entrepreneur. She was well aware of the double bind her desire to keep the divorce amicable placed her in. "I always knew that if I had said, 'I'm going to fight,' or threatened to take him to court where his assets would be public, he would have given me oodles and oodles of money," Winifred concedes frankly, "but I didn't want to do that because I wanted to be friends with him and his children." In retrospect she feels that decision was rooted in her own low self-esteem, saying, "He would have gotten angry, he would have screamed and yelled, but he would have gotten over it. In fact, he would have respected me more if I had asked for more money."

Winifred interviewed two lawyers and hired the one who urged her to educate herself financially and really figure out how much her lifestyle would cost. "She basically started an empowerment process for me," says Winifred, who went from complete ignorance of money matters to shrewd financial manager in less than a year. The divorce took six weeks and cost Winifred $2,000. Though the settlement she obtained was much less than she was legally entitled to, it was reasonable: $60,000 a year for four years, and an investment property that yields about $30,000 annually. Marty and she have remained friends, and when, six years later, she asked him to co-sign a loan so she could buy an apartment, Marty changed the settlement and gave her the money outright. One reason the arrangement was attractive to Marty is that it perpetuated the pattern established in the marriage, during which the wealthy businessman doled out spending money—plenty of it, but only on request—to his dependent wife.

In most cases the effort to stay on good terms had no financial payoff, but the psychological benefits were worth it. Feeling as though you've done your best to be civil and kind has its own rewards. Many, many women expressed their relief that their marriages hadn't ended in total alienation, that they and their ex-husbands could still count on each other, that they had conducted themselves in a civilized manner and maintained their integrity. It would be hard to put a value on those feelings. Whether or not it made sense legally or strategically, behaving well to maintain a connection has authentic value. The transition from wife to ex-wife may be bearable, but that from friend to enemy often is not.

The Cons of Being Conciliatory

No one wants the horror stories, and everyone has heard them: court battles that drag on for years, houses sold to pay legal fees, children torn limb from limb. Nancy started out "telling lawyers, 'I don't want this to be nasty, I want you to be nice,'" but her conclusion was a different one: "Boy, was I dumb." Trying to keep things amicable can make a divorce take longer, cost more, and earn a woman no goodwill from her husband anyway.

Amanda's divorce was a costly nightmare. A quiet woman with fingers inkstained from her hours at the drafting table and cheekbones you could hang a hat on, she had been verbally abused by her much older, emotionally unstable husband throughout their four-year marriage. Anton never hired a lawyer and his behavior grew even nastier and less rational after his architect-wife moved out with their toddler son. Yet Amanda felt badly about separating father and child and worried that Anton would have a hard time getting by without her meager but steady income. Looking back, Amanda no longer feels guilty, but rather that "I was far too compassionate. I felt too much responsibility for his feelings."

Still in graduate school at the time, Amanda waited nine or ten months before consulting a lawyer, and when she did, she wasn't prepared to roll up her sleeves and do battle.

I went to a really nasty bulldog of a lawyer and he scared me. He scared me because I don't think I was ready to hear the nastiness. The whole thing was just so degrading that I walked out and thought, "I can't deal with this guy." I should have just stuck with him and gotten it done, in a month, whatever it would have taken. Instead I went to a law firm that was more genteel, and they ripped me off for thousands and thousands of dollars.

Her fancy lawyers dragged things out, and her husband obstructed the procedure in every possible way. Two years and

$60,000 later—her father, an investment banker, paid—she finally obtained what she was after: an order of protection and supervised visitation.

Anton has never paid a penny of child support, but Amanda never went after him. "I just couldn't face talking to another lawyer about it," she explains wearily. Besides, he doesn't have any money and she's not eager for any further dealings except where it concerns his occasional visits with Evan. In retrospect, her advice is unequivocal: "Demand everything, everything that you think you might need or want in the future, the long term, and you won't feel bitter later about undercutting yourself. Go after everything that you deserve."

When a spouse is flat-out determined to obstruct the divorce, whether for profit or revenge, there's a limit to what smart lawyers and good intentions can achieve. Such was the case with Mike, whose wealthy family continued to insulate him from his responsibilities. "Mike always came out smelling like a rose," Ginny says ruefully of her eternally boyish husband, whose three-day drinking binge when Ginny gave birth to their third child was the beginning of the end. "He had time, he had money, and he didn't have the care of his kids." Ginny consulted a lawyer who was supposed to be absolutely the best. Like Amanda, she wasn't ready to hear what the lawyer had to say:

> She was very frank. She was so harsh, though, that it was very intimidating to me, although the things that she was saying weren't intimidating. I made a comment like, "Mike will not cooperate," and she goes, "I don't know what your problem is. You're obviously the strong one here. What are you afraid of?" What she was saying was right, but it was in a manner that I wasn't strong enough to handle. She was like a barracuda, probably really good, but I was not prepared to be that barracuda too.

Ginny says she did turn into that barracuda later on, though, because Mike fought it every inch of the way, "because in order to keep it moving ahead, I had to pay for it. All these different searches and transfers, I had to pay for all of them. He just sat there and refused to do anything." It took over two years, and cost Ginny $10,000 in lawyer's fees.

Whether it's the woman or her attorney who gets tough, sometimes that's what's called for to wrap things up. Over the five years
that have passed since her divorce, Yolanda, the PR director, has
figured out how to handle it when issues crop up around Will, their
learning-disabled eleven-year-old. Never conciliatory, she now calls
Steve's bluff instead of trying to negotiate. When she changed jobs
not long ago, Steve called up and said, "You must have gotten a
nice raise."

> I said, "That's right, I did." He said, "Then we'll be cutting back on
> the child support, won't we?" And I said, "No, I don't think so." He
> said, "Oh yes we will, there's no court in the country ... "—his
> "expert" number—and I said, "Okay, you check with your lawyer,
> and if he wants to do something about that then your lawyer will get
> in touch with my lawyer, and we don't have to talk about this." He
> blustered for a while, and I guess eventually he did talk to his lawyer,
> because that was the last I heard of that.

Distasteful though it may be, sometimes a woman has to get
ruthless. But there are no clear-cut rules as to just how or just
when. Depending on how long things have gone on, how intransi-
gent people are being, and what she stands to gain or lose, it's a
judgment call.

Divorce on a Shoestring

All else being equal, the client with the deeper pockets has an indis-
putable advantage in any legal proceeding. This is seldom the wife,
thanks to the wage gap: American women earn 72 cents for every
dollar earned by men. Even in this small sample only one woman
out of five earned as much as or more than her spouse; most husbands made much more than their wives did.

Self-possessed in a tailored gray silk pantsuit, her auburn hair in
a severe bowl cut, Dina herself is an attorney, though her specialty
is immigration law. She saved money by doing a lot of the paper-
work herself and consulting her lawyer only for major decisions, so
the divorce cost around $1,000. The mother of two boys, Dina

nevertheless managed to get some really bad advice. "My lawyer didn't want me to put anything in there regarding college costs because he didn't want *me* to be responsible for half those costs. I can't believe now that I let him do that, because I'll probably end up paying the entire amount," she says with a groan. The other mistake was not cutting up the credit cards she shared with her husband. Their verbal agreement that each would assume responsibility for his or her own bills fell apart when Alan went bankrupt and the court automatically assigned Dina half the debt.

Women tend to be handicapped by ignorance of money matters. A startling number of the women I interviewed had no idea how much their husbands made. Some, like Yolanda, whose husband ran up a $300,000 gambling debt—"quite the eye-opener *there*," she quips—were deliberately kept in the dark by their husbands when it came to family finances. Others looked the other way, leaving money management to their mates even in the face of blatant incompetence, and rationalizing the arrangement as appropriate or "natural."

This attitude leaves women very exposed. They may be too cash-poor to plunk down a retainer for a good lawyer. Few lawyers are willing to take on a client who can't pay them until money controlled by her husband has been freed up. Also, under community property laws, to claim their share of the marital assets women may have to prove what is rightfully theirs, because it's not spelled out. This means she must know what a couple owns and in whose name the credit cards, deeds, car, and the like are held. If only the husband has the information, his wife is at a real disadvantage, especially since the courts frequently refuse to force disclosure from the party with the facts. And if assets are concealed, it's much easier to falsify information. Such women are vulnerable to the cash-stashing tricks of husbands like Bernadine's in *Waiting to Exhale*; they are also less adept at the financial rearrangements divorce requires, and more daunted by the prospect of financial self-sufficiency.

The old adage that "you get what you pay for" often rings true when a woman's funds are limited and she needs to hire a lawyer. Laurie's story shows what can happen when a woman lacks both money and foresight.

"Beg, borrow, or steal, but get the best lawyer you can get."

Laurie, now a homemaker in suburban Washington, D.C., spent much of her marriage putting Mel through dental school, and had no savings at all. At thirty, after nine years with "Doom & Gloom," as she nicknamed him, Laurie decided she'd had enough. The first counselor she consulted, instead of advising her on how to get out, said, "'After all these years of putting him through school, can't you work it out?' She felt that I was nuts." Another therapist put the consequences bluntly. "She said, 'Look, you're thirty years old, you have a child, you may never be in another relationship. Not too many men out there are interested in a ready-made family. If you want to end this marriage, that's a risk you're going to have to take.'" It still sounded better to Laurie, who leans against a refrigerator plastered with children's artwork, dressed in a gray sweatsuit and Reeboks.

Doing it cheaply, she found, definitely has its drawbacks. Since Laurie had no money, a girlfriend who worked for a lawyer—who was subsequently disbarred—drew up a template for free. Mel gave Laurie no money at all during the divorce proceedings, and, says Laurie, "I got a lousy deal. I got $50 a week child support for the next twelve years." Nor, despite her investment in Mel's education, did she get alimony. Her advice is straightforward: "Beg, borrow, or steal, but get the best lawyer you can get. Make your settlement early on. Don't think that you're going to go back at some future date and change everything. If you have to borrow money from your family, do it, and get everything in writing that you can think of. Get the most experienced person you can find. It's expensive, but you're going to live with this agreement."

"She said, 'Let's try to do this so that he won't get a lawyer.'"

By doing their own legwork, researching local resources, and other clever strategies, many women are able to get the most bang for their buck. Helen, who lived in a trailer on the outskirts of Tallahassee, knew that lawyers were expensive and didn't know how she was going to afford one. "I can't remember how I heard about it, but I found a women's legal aid society. I got a good lawyer, at a very reasonable price," she recalls. The divorce cost approximately $450, and papers were signed in thirteen months. Helen's husband, a passive-aggressive man slipping deeper into alcoholism, never hired an attorney, and Helen credits her own lawyer with some excellent advice on that point. She said, "Let's try to do this so that he won't get a lawyer. You're working and he's not and you could end up paying him alimony." That made sense to Helen, who gave ground on a number of points, including Ron's pie-in-the-sky plan to buy her share of the trailer. "But I agreed with the stipulation that if he didn't come up with the cash, it would immediately be put on the market," she explains, "and that's exactly what happened."

Like Helen's and Amanda's husbands, a number of men never hired counsel, forgoing legal representation out of pure orneriness or because they couldn't afford a lawyer either. Undoubtedly these men ended up with a smaller piece of the pie, but they were less hungry or there wasn't much pie to start with.

Letting Him Take Care of It—"I was dead sure I wasn't going to spend another penny on that guy."

For those whose circumstances or temperaments don't require a speedy legal resolution, waiting for him to take the first legal steps can work, and it sure is cost-effective, as three women I interviewed found out. Beverley, a middle-class minister's wife with one young son, had been planning to use a do-it-yourself book when papers arrived from her husband stating, falsely, that she had been served. "The agreement's legal unless I decide to contest it," she says with a shrug, "and I didn't want to." She and Percy had already decided on a custody arrangement, and all Beverley wanted was the $150 a month in child support her husband had agreed to and which his parents paid until she moved out of state six months later. (She ended up suing for back child support payments, and now his second wife sends the check.)

Two of the women I interviewed were completely broke when they walked out on their husbands, and decided to let the men go through the trouble and expense of drawing up papers. As Bibi put it, "I was dead sure that I wasn't going to spend another penny on that guy." As long as there is no property to settle and the woman is in no hurry to remarry, in some cases it's possible to file for custody and child support even if the couple is still legally married. This wait-and-see strategy works especially well if the ex-husband ends up with a pressing reason to get unhitched legally, such as an impatient girlfriend in the wings. A number of husbands of women in this survey did end up assuming the financial burden—usually small, because these were not divorces with a great deal to be contested—even though their wives had made the first move.

Spelling Things Out—Different Kinds of Agreements

Divorce agreements come in all sizes. My agreement with my ex-husband is eighty-seven pages long, and sets out every imaginable aspect of property division, custody, and child support; it works for us. Much as I chafe under the rigidity of the custody schedule, I love being able to plan ahead with certainty. My significant other and his ex-wife, on the other hand, don't seem to be able to look more than a week ahead and are constantly changing things around at the last minute. Of course this drives me nuts, but then again he thinks the constraints I live with are ridiculous. Their two-page agreement stipulates little besides the fact that custody and child-related responsibilities are to be shared fifty-fifty. This means that everything has to be negotiated; it works for them. Ginny, the ice skater with the alcoholic ex, looks askance at a friend's one-week-on, one-week-off custody arrangement because it's so structured and inflexible. Her arrangement with her ex-husband, Mike, is very loose, which means that "if he goes out with his girlfriend, the kids are with me," she observes wryly. "Now I know why they make those arrangements—so that you can have a life." The best agreements fit both the resources and the dynamic of the couple, though that can be hard for all parties to assess in the throes of divorce.

"I wasn't in a state to think about [paying for] college, insurance, mechanisms for enforcement . . . "

Especially when children are involved, distraught women can be desperate to work out an agreement as fast as possible. Emotional distress can lead them to agree to a less-than-satisfactory arrangement. Had I not found sharing the house with my not-yet-ex so intolerable, I would certainly have been in better condition to follow my lawyer's advice to hold out for better terms. Virginia, the Sissy Spacek look-alike and cartographer, found herself pregnant at twenty-six, married the father, a self-styled playwright, on "a wing and a prayer" and moved to Grand Rapids. As soon as Ian was born, she encountered a panicked, jealous, irresponsible man she'd never met before, which was "just scary, just scary as hell," Virginia says. Four years later she threw in the towel. She was able to relocate to another apartment in the same building (her address didn't change, so her custody claim wasn't affected) but she badly wanted to get things over with. Along with being in a hurry, her other mistake was that she "wanted to be fair" without having a clear idea of what would be fair to both her husband Christopher and herself down the line. Consequently she's hostage to an agreement that's too short, too vague, and grossly inadequate.

Custody of Ian was the only thing that really mattered to her, and Christopher backed down from his threat of a custody fight when Virginia made it clear she'd go to court if necessary. Her divorce cost $500 and took about four months to go through, because he didn't contest it; they actually signed the papers on the kitchen table. "It was very cheap, and I have the world's worst divorce agreement," says Virginia. Unfortunately she's been living with the consequences ever since.

The $200 a month Virginia gets for child support was supposed to be reevaluated two years later, but Christopher refused to renegotiate. "It's unenforceable," she says grimly, "and it costs more money to change it than I can afford." Virginia wishes the agreement were "*much* bigger, much more comprehensive. I wasn't in a state to think about [paying for] college, insurance, mechanisms for enforcement. I was still under this delusion that there would be some level of cooperation." Christopher has refused to speak to her since she left Michigan five years ago, though Ian visits his father regularly.

"There is no dignity when you have no home and no money."

Amy, conversely, came away from her child support negotiations with the opposite conclusion from Virginia's: she desperately wishes her agreement were *less* comprehensive. She started out with a lawyer who, after the proceedings had dragged on for over a year, said, "'You know, this is really crazy, this is really tearing you apart. Just give up the house, give up everything, walk away. Just keep your dignity.' Bullshit," says Amy, tugging angrily at the sleeve of her oversize cardigan. "There is no dignity when you have no home and no money. He told me to leave, and I began to sign a document that was going to just put me out on the street with my daughter." Fortunately Amy decided to get a second opinion from another lawyer, who reviewed the agreement and said, "'Why would you do that?' It was that simple: 'Why would you do that?' The answer was, because I was in pain for two years. I couldn't stand it one more second, I thought it would be better to just walk away."

A sturdy woman whose high cheekbones and black eyes hint at the Cherokee blood on her father's side, Amy refused to move herself and their four-year-old daughter, Cara, out of their rambling house in Oakland. Her husband too refused to move, or to commit to a child support schedule. Eventually Amy ran out of both money and stamina. "Martin's lawyer, who really did not like women, pushed all the right buttons. If I showed any emotion, this man would stand up and say, 'That's it! You shut your mouth, little girl, or we're walking out of here.' He was a really bad, hard-assed person, and he did a real disservice to me and Martin. He got a much better agreement for Martin than I got for myself, because they could go on forever, and because there was just so long I could go on being attacked." She and her husband finally consulted a family therapist, who kept saying, "Your agreement's got too much in it." But Martin and his lawyer "would *not* make it more flexible," Amy says angrily. She particularly resents a stipulation that forbids her from moving more than seventy-five miles without jeopardizing custody, though Martin had to sign it too.

More vulnerable than her husband financially and emotionally, which Martin and his lawyer took keen advantage of, Amy eventually could not hold out any longer. Though heartless, the holding out strategy is an effective one for the partner with more resources. The house is most couples' largest asset, so attorneys advise warring spouses not to jeopardize their claim by moving out. This takes a terrible toll on everyone involved, especially children exposed every day to parental tension and anger.

In retrospect Amy thinks going to court right at the start would have saved her money. As it was she spent $15,000 on an agreement that could have been much simpler, and its comprehensiveness did not protect her from a major loophole. Martin finally agreed to pay child support based on a percentage of his income, an agreement he then neatly circumvented. Her ex-husband lives off an inheritance that was in his mother's name and which Amy was unaware of. (Full disclosure of his assets would have protected her from this costly mistake.) During the marriage Martin made money as a cabinet-maker and carpenter, and Amy unwisely but understandably assumed this income would continue. "He paid no taxes last year, let me just put it that way," Amy explains tartly, "and when you do the agreement, 17 percent of nothing is nothing. So I got stuck." She also suffers from lupus, which is both fatiguing and expensive. Not bitter at the time, Amy describes her present financial situation as "a real hardship. But I'll tell you the truth," she adds. "I feel it's worth it."

Foreseeing that loophole because her alcoholic husband wasn't working at the time of the divorce, Ginny settled instead for a flat $400 a month in child support for her three children. Needless to say, she can spend that much in one trip to the grocery store, but Mike does pay it, and Ginny makes decent money as an art director for a Rochester home-furnishings company. Negotiating a settlement is complicated, and where children come into play, mistakes and misconceptions cast a very long shadow over the years. Amy makes a case for holding off on a final arrangement, and the idea of a periodic review of custody and support arrangements does make sense. "I don't think anyone should have a custody agreement until you've lived apart for a while, to see how you feel and how your child feels. Then go back and make an agreement," she suggests. "How the hell do you know what's good for your child in three years, or two years, or one?" Long or short, the essential ingredient of a divorce agreement when children are involved is flexibility in the long haul.

Custody and Child Support

What works for one family is anathema to another, and truly putting the child's interest first requires maturity and a generous spirit. The big effort is to envision how decisions made now will affect children in the long run, and to acknowledge that arrangements may need to change over time. Parents move. Plants close. People remarry. Children turn into teenagers with lives and logistics of their own. Flexibility can't be mandated on paper, any more than goodwill can, even when it's the child who suffers.

Custody issues far outweighed any others during the divorce proceedings for many women in this study. This was certainly true for me; I was haunted by a recurring vision of my children falling through the air like rag dolls. In the crudest sense children are

THE BASICS OF CUSTODY ARRANGEMENTS

A custody arrangement has a number of elements, which are described in various, sometimes overlapping, terms. They include: who will the primary caregiver be and whose home will be the child's primary residence; whether decisions regarding the child will be made jointly or by the primary caregiver; who has the right to information about the child; and what the visitation arrangement will be. When one parent has sole custody, he or she is designated the primary caregiver and the child lives with that parent, who has the final say in decisions regarding the child, but should consult with the other parent. When joint custody is awarded, parents equally share information and decision-making regarding the child's health, education, religion, and activities. Joint custody does not necessarily mean that a child's time is divided equally between two households; it's common for one parent to have the primary residence. Joint custody is seldom awarded by the court unless both parents agree to the arrangement. Legal custody means the authority to make decisions about a child, and physical custody is control over a child's actual whereabouts. Visitation is a noncustodial parent's right to see the child regularly. (The long-term pros and cons of different arrangements are discussed in Chapter Six.)

property, incalculably precious ballast in the equation of divorce. They're also expensive, and being hard up makes it harder to provide everything for a child that a parent might wish to. It's tempting to make financial concessions when so much is at stake, but it can be disastrous in the long run. Few responsible parents can afford to allow their emotional need for custody to overshadow their material requirements. Custody and child support are inextricably entwined, a fact cruelly and ironically brought home to mothers forced by divorce into working full-time—and whose claim to custody may be jeopardized by that very employment.

THE BASICS OF CHILD SUPPORT

Child support is money paid by the noncustodial parent to assist with the expenses of raising children, until they reach age eighteen. In New York State, children may sue for support to age twenty-one. Most child-support guidelines are based in large part on a percentage of the income of the parent paying support, adjusted for the number of children. The nationwide Child Support Standards Act states that 17 percent of the first $80,000 in income of the noncustodial parent should go to child support. Medical insurance and school tuitions are not automatically provided for and must be negotiated separately. Working out a support arrangement that's equitable to both partners and adequate for the children is a complicated affair. How many children require support? What about children from other relationships? Does the child have special needs? What are the net incomes of the parents (and how is net defined), and what weight should be given to the income of the custodial parent? How will custodial time be shared, and what credit should be allowed for extended visitation and unusual custody arrangements? How should health care and day care costs be allocated? Is there a way to accommodate changes in income and lifestyle as time passes? There will always be parents who can't or won't come up with the bucks, but they're far more likely to do so if they think the deal is fair.

DETERMINING THE "BEST INTERESTS OF THE CHILD"

The law in most states says that custody should be awarded to either or both parents based on the "best interests of the child." It sounds good, but determining what those best interests consist of, especially over time, is by far the most difficult and emotionally wrenching element of a divorce agreement. They may differ radically from family to family, and even from child to child. Even parents who adore their children don't always agree on what's best for them, even when they're trying hard to act in tandem. Despite bitter legal wrangling, both Amy and Martin were almost obsessively committed to keeping everything in their Oakland home unchanged for their four-year-old daughter, right down to family vacations and Sunday breakfasts together. "We really kept up a big front for a long time, and I think Cara bore the brunt of that, all the tension," says her mother ruefully, though it seemed the right thing to do at the time.

To my mind my husband's moving back in with me and the children clearly demonstrated that he was not capable of acting in the kids' best interest. "Let's leave your and my ideas of 'best interests' out of it," he proposed through clenched teeth. Creepy though it sounded, it was a sensible proposition because it lowered the emotional temperature, which *was* in the kids' best interest. At the time, we were far too caught up in our separate dramas of loss, panic, and control to look at the big picture. In due course, with the help of our attorneys, we pretty much managed it, though we need the occasional tune-up at the hands of the family therapist built into our agreement.

Attorneys don't always have a clear fix on what's best for the child either, especially when the child's and the client's interests conflict. Monica divorced in 1975 when custody was almost invariably given to mothers, but just to make sure, Monica's lawyer suggested, "Why don't you pretend [your husband] hit you? Why don't you not feed him? Why don't you force him out?" Monica's reply was to point out that a five-year-old child would be witness to whatever she did. "I'm not going to have his mother behave like that," she said firmly. "It'll have a lasting impression on him."

If custody cases go to court, family law judges rule on a case-by-case basis and have great latitude in interpreting the "best interests of the child" standard. Gender bias—whether pro-mother or pro-father—may play as important a role as other, legitimate factors in determining a family's destiny. It's frightening to think that the welfare of children should be so hostage to prejudice or whim. Typically a lawyer, paid for by both parties, is appointed to represent the children, and a psychiatrist performs a family evaluation. An excellent resource in determining what is genuinely best for children is *Parent vs. Parent: How You and Your Child Can Survive the Custody Battle*, by Stephen P. Herman, M.D. (New York: Pantheon, 1990). In an ideal world, custody and support arrangements would be subject to periodic review and revised if necessary to suit the ongoing needs of the children.

Who Gets Custody and Why

Custody is still overwhelmingly, though not automatically, awarded to mothers. This is partly because they ask for it more than fathers do, and partly because legal presumptions and traditions favor the mother as custodial parent, especially when the children are small. Two-thirds of the women I interviewed have sole custody, and one-third share joint custody. One father obtained sole custody when his emotionally disturbed wife deserted the family for three months, and another woman, a manic depressive, chose to give her ex-husband physical custody of their two young daughters.

Activist dads are fighting the gender bias that assumes women should automatically get the kids, and many speak angrily of lack of access and loss of control over their children's lives. Joint custody is on the increase, in part because of campaigns being waged by a number of fathers' rights organizations whose members want a role in their children's lives beyond that of the guy who sends a monthly check. No one benefits from a blind bias toward maternal custody: it sets the stage for dads to be left out of the loop, and pressures mothers who might otherwise decline primary custody not to do so, even when fathers are better suited to take it on. It also has serious financial consequences, since the more regularly dads see their children, the more likely they are to pay child support.

WHEN FATHERS SUE

Few words are more terrifying to mothers than "I'm suing you for custody." Fathers may pursue custody because it can be cheaper to keep the kids—an ironic effect of stricter laws requiring the non-custodial parent to pay support. With increasing frequency fathers are being counseled by their lawyers to get custody from their wives, and the child support that comes along with it, as an economy move. Many fathers genuinely want custody for emotional reasons—to be the primary caregiver—rather than economic ones. Others, however, use the threat of a custody fight as a bargaining chip in return for financial and logistical concessions, a tactic lawyers are not hesitant to recommend. Ploy or not, it's an effective one, because of the sheer terror it invokes, because judges don't necessarily see it as bogus, and because of the current climate in the courts.

Twenty-five percent of the mothers I interviewed were sued for custody and an additional 20 percent of the fathers threatened to sue. None was successful, but nationwide, according to studies of gender bias in the courts, the best evidence is that when fathers challenge mothers for custody, they obtain it 70 percent of the time. This can be attributed to a shift in the states' attitudes toward custody in the 1970s, when the "maternal preference" rule came to be seen as discriminatory to men. Many states passed joint custody laws and gender-neutral standards that in theory required that each custody case be individually evaluated.

These changes came about because the women's movement had inspired people to question traditional sex roles and the assumption that fathers are less qualified than mothers to be primary parents. The movie *Kramer vs. Kramer* poignantly captured this new awareness. During his big moment in court, Mr. Kramer, portrayed by Dustin Hoffman, stands up and asks, ". . . what law is it that says a woman is a better parent simply because of her sex?" The rhetoric is touching, but Kramer doesn't even know how old his child is. In real life, Woody Allen had never met his son's teacher and couldn't name the family pediatrician, but that didn't stop him from suing for custody. The underdog-friendly sentiment that Johnny-come-latelys deserve an equal shot at parenting is a trendy one. But if these parents were so concerned about their children, how come they abdicated responsibility while it was theirs for the taking?

When it comes to a custody battle, mothers of small children may be damned if they work and damned if they don't. If they stay at home to care for their children, judges may rule that they lack the financial resources to provide adequately for their children. Mothers who work full-time, on the other hand, may be held to be neglecting their latch-key or day-cared children. Those were the grounds on which Marcia Clark's ex-husband sued. "Her prosecution of O.J. Simpson has made her one of the best-known working mothers in America," wrote the *New York Post,* "and now it has landed Marcia Clark in a bitter custody battle with her estranged husband, who contends that her grueling workload is harming their two sons." Jennifer Ireland, a twenty-year-old Michigan resident, was sued for custody because she planned to put the child in day care while she got her college degree. The father's plan was to leave the child with his mother (the child's grandmother) while he himself worked. On this basis, the judge ruled in his favor, though the ruling was overturned by a higher court a year later, in 1995.

Men face no such double standard. Their full-time jobs are evidence that they can provide better for children than a mother who earns less, although few fathers are willing to throw in the professional towel and stay home with their offspring. "Men who win custody do not settle into new roles as primary caretakers of their children," observes legal activist Karen Winner. "They remain fathers and assign to the new females in their lives—their mothers or their girlfriends—the primary duties of raising their children." Nor do husbands of working women compensate for parental shortfall by spending proportionately more time with the children themselves.

Sole Custody

Parents who seek sole custody are often perceived as cruel or vindictive, motivated by the desire to deprive the spouse of his or her child. In fact women often seek sole custody with the welfare of the family foremost in mind. Since her divorce Karen has actively encouraged the relationship between her son and his father, but she felt that as the only "grown-up" in the family she had to be able to set limits and enforce them. Sole custody gave her that clout, which was crucial in breaking destructive patterns affecting all three people in the family. Her marital problems had surfaced when Ivan was four, and Karen, already struggling to keep the household going, refused to consider having another child. Neil's response was to start looking for a woman who *would* oblige him in the baby department, and he began a series of "friendships." Seeing him as an "innocent baby flailing about," Karen put up with the situation until it became clear he simply wasn't going to move out. She finally consulted a lawyer, who instantly told Karen, "This is a custody thing. I think he's probably even had some advice that if he moves out and has these girlfriends he'll never have a chance." It was all true, a complete bombshell to Karen.

Karen borrowed heavily to pay for "a really expensive woman lawyer who was on the state board of marital lawyers, knew the judge, and really wielded that authority." The divorce cost over $50,000, leaving Karen, a sculptor, seriously in debt, but "I'm really glad that I spent all that money and all that heartache getting the sole custody of Ivan. It was painful and costly, but I think that Neil has a different relationship to me as a result, and a more respectful relationship to his son as a result," she maintains. "It's been right for all of us." Karen describes the support that Neil pays as "just barely adequate. But what my lawyer and I finally decided was that it was better for him to have an amount that he wanted to pay—that he *would* pay—than an amount that he would never pay. And have nothing, and an ongoing battle."

Sole custody gives that parent considerable control over the child's exposure to unsafe or uncertain circumstances. That was the motivating force behind Nancy's successful battle for sole custody of their three-year-old. The tableware designer was certain that Eric was far too disorganized and irresponsible to take care of Lauren for extended periods of time. Weekends were all she thought he could handle, and indeed Eric often doesn't show up at visitation times, one week night for dinner and every other weekend. "I'm always having to cover his time with baby-sitters," says Nancy. "He has never, ever, ever asked for more time" with Lauren. Some parents just aren't equipped to raise children single-handedly, though mothers are demonized if they admit as much and yield custody. Another factor in Nancy's case was her certainty that in Eric's mind the fight had nothing to do with Lauren and everything to do with punishing his wife.

Joint Custody

Joint custody was devised to replace the traditional model, in which the mother got sole custody and the father visitation rights. It has clear advantages: children see both parents regularly, and two people share equally in the enormous responsibilities of parenting. Among the findings of Judith Wallerstein's Children of Divorce project, a six-year study of sixty California families, was that without the legal right to share in decision-making, many parents withdrew in grief and frustration, and children experienced this as rejection. The disadvantages of joint custody are equally clear: kids are often shuttled between two homes; an enormous amount of flexibility and communication is required; and exposure to unwholesome circumstances cannot always be controlled. When cooperation breaks down, kids are directly exposed to the hostility between the mother and father, which can be devastating. Some families aren't temperamentally suited for the back-and-forth, but others can make it work.

My husband fought hard for joint custody, and only in retrospect did I understand the depth of his panic that I could take his children away from him simply because I was their mother. The basic deal he and I brokered, alone at our dining room table two months into our negotiations, was that I would agree to joint custody if I got the children four days a week and if he agreed to move out. That's what happened. Though I suspect that it makes all four of us feel unsettled at times, the arrangement is a good one for our family. I hope he will become less rigid as the children enter adolescence, but I have no power to do anything but hope.

Holidays alternate yearly, and vacations are split fifty-fifty. The hassles of having two homes takes its toll (the raincoats are inevitably at his house on rainy mornings, and if they had two sets, they'd both be there). But he lives only two blocks away, which makes the arrangement far more workable than if he lived across town or in a different school district. The kids have keys to both places, so if the Rollerblades are at one house and the helmets at the other, it's not a crisis. Switches routinely occur on the neutral turf of the schoolyard (one parent dropping them off, the other picking them up), so our children don't feel torn in two. The important stuff—recitals, school tours, teacher conferences—my ex-husband and I attend together.

Though my ex-husband was always a loving father, I'm certain he is a more aware and active one than if we had stayed married, when passing on tedious aspects of the children's day-to-day care was a way of exerting his authority over me. The kids benefit from the arrangement, though there are moments when I'd love the sovereignty of sole custody, to be able to schedule appointments or change schools without having to negotiate. But it gets easier as the children get older and better at saying what they need. Selfishly I wish my ex-husband would let go more, but I know that if my children's worst post-divorce problem is that both parents are hungry for time with them, they're lucky. It would be far worse to have to explain a father's absence or disinterest, as many mothers do.

Splitting Up the Kids

The kind of intensive communication and back-and-forth that joint custody calls for was out of the question in Sophia's divorce. Mac had always been more involved with his oldest child, the only boy, than with his two daughters. "He used to take my son to visit his girlfriends," offers Sophia, the woman raised by migrant workers, by way of an example. Splitting up the kids was her solution to the custody quandary. Sophia, who was raped by her husband when she refused to reconcile for a third time, contacted an attorney when she found out she was pregnant again. In California in 1966 it was necessary to wait out a year of separation before going before a judge, and in any case Sophia wanted to wait till the baby was born to make sure the child hadn't been harmed by the violence of her conception. "I couldn't cope with anything else [except the pregnancy]," she says simply. Her son, then nine, never wanted to stay with his mother, and at the trial, the judge asked to talk to the boy alone in his chambers. "He came back out and said to me, 'Your son wants to go live with his father. He will be nothing but trouble for you and your two daughters. You three have a life to lead. I recommend that you let him go.'" Sophia did, and has never regretted it.

Sophia was awarded $1 a month alimony to establish a claim if her ex-husband ever came into money, and child support of $100 or $125 a month for her two daughters—she doesn't remember the exact amount because she never received a single payment. Sophia never pursued the matter, however, because she wanted as few dealings with Mac as possible. It was rough, recalls Sophia, but she and her girls were happy. When she remarried two years later, her second husband wanted to adopt her daughters, then three and five. "The only way to get my ex-husband to say yes was by agreeing to take him off the books," says Sophia. "He said, 'Don't nail me for back child support or alimony and I'll sign the adoption papers.'" It was a deal.

Whatever their economic status, mothers who divorce face tough double binds. If well-to-do, they may be pressured to give ex-husbands a disproportionate share of their assets to ensure custody or to settle quickly. If poor, they face the exhaustion of making ends meet and the stress of custody threats from ex-husbands like Violet's, who at the time was providing state-subsidized child care in her home. When William would drop their sons off at Violet's housing project after his occasional visits, he would taunt her.

He would say I wasn't taking care of the children right, that he would come and take the children away from me through the court. It just made me stronger, because I had made up my mind. I could see what road he was on, I said, "My poor babies. He'll never, *never* get my children." That's when I got my inner strength.

Divorce strengthens rather than weakens a woman's tie to her children. It requires that she define and protect the job of being a mother, and children fought for are never taken for granted.

Leaving the Paperwork Behind

It's difficult to generalize about women's experience within the judicial system because circumstances and personalities are so diverse. Every lifesaving piece of advice seems to have its counterpart in counsel happily ignored. Many women just want out regardless of the price and penalties, which is more a testament to the misery of their marriages than an endorsement of the future that awaits the divorced woman. Many lack both the know-how and the desire to obtain and enforce a genuinely equitable agreement.

But even those women I interviewed who earned significantly less than their husbands focused during their divorce process far more on how to survive on their own than on "taking him for all he's worth." Most husbands behaved decently, if tightfistedly, once they had accepted the inevitability of the divorce. The stereotype of the vindictive wife was not borne out, and *War of the Roses*, no-holds-barred lawsuits were the great exception as well. Once papers are signed, women put hearings and subpoenas behind them and turn their energies to shaping their futures.

CHAPTER 4

Money
and
Work

$\mathcal{A}$s women think about ending their marriages, their biggest worries are usually financial. Will they be able to take care of themselves? Provide for the children? Live in a decent place? Get out of the house every so often? Some postpone their divorces until their careers are established or they have some sort of savings. Others cut and run and suffer the financial consequences, and these can be significant.

How Divorce Affects Women's Standards of Living

When Lenore Weitzman's *The Divorce Revolution* was published in 1985, the media went to town with her findings that in the wake of divorce men's standards of living improved by 73 percent and women's declined by 42 percent. These numbers are perennially seized upon to demonstrate how divorce in general, and no-fault laws in particular, impoverish women. This was in spite of the fact that no researcher was able to replicate Weitzman's findings, which Dr. Weitzman herself conceded in 1996 were "probably incorrect." Current research shows a 20 to 30 percent drop in women's standards of living and a 10 to 15 percent improvement for men's.

Though these figures are less drastic, the inescapable fact is that women who divorce do suffer serious financial consequences. Statistics abound: only 15 percent of divorcing women are awarded alimony or maintenance, and only 6 percent of the women in this survey. Thirty-eight percent of divorced or separated women with kids live in poverty, compared to 17 percent of men with custody, according to the 1994 census. The demographically "average" American woman is raising children on $16,000 a year.

Divorcing women are hardest hit when they have children; many really have to scramble, often for years. Mothers generally receive primary custody, and as noted in the preceding chapter, child support is often inadequate or nonexistent. Amy, one of the women I interviewed, is typical of a woman who is financially strapped. She is struggling with the cost of treating her lupus, shouldering expenses that had once been split, and making up for her ex-husband's evasion of child support payments. The family business she works for has been going through hard times, and Amy works only part-time so she can care for seven-year-old Cara after school. Nevertheless, despite her bleak financial picture, she says firmly, "I feel I have a certain peace of mind that I did not have before. I could *not* go back to that."

Amy's attitude is the rule, not the exception among the women in this study. Not having to beg for money, account for it, or ask permission to spend it made the tradeoff worthwhile for every single one of them. A decent salary was a giant help, but lack of money was *not* reason enough for these women to stay married. One after another described living on less but having it feel like more. Working women without young children suffered only slightly, but on average it took a few years for women to get back on their feet, and a few needed seven or eight years. Remarriage was the passport to financial security for approximately one-third of them. The rest learned to budget and manage their money, completed their educations, got better jobs, worked harder than they ever had before, and proudly achieved self-sufficiency.

"I thought I was better off on my own, and yet I think I paid double the rent," says Esther, who moved out at age fifty-four, after twenty-seven years of marriage, and now lives in West Palm Beach. This is a woman without a romantic bone in her body, a book-keeper who handled a big company payroll and managed her family finances down to the nickel. With a shrug she admits that the math doesn't justify her sense of material well-being. Her rationale for her positive feelings is that she is "so much more in control of what I'm doing, although I made decisions in the marriage too. It's a whole different thing." The exchange of prosperity for autonomy was well worth it.

One reason that women feel better off despite reduced circumstances is that they are less likely than men to measure their well-being materially. Certainly women find sustenance in friends and children that is beyond price.

As noted in the Preface, Weitzman herself observed that even older women who felt economically deprived and financially "cheated" by their divorces felt personally better off. To explain the apparent dichotomy between how much money women have and how they feel, it's necessary to look at the complex relationship between work and identity, and income and power. This, in turn, requires a look at the underlying economics.

The Roots of Economic Discrimination Against Women

The principle of the family wage—that a man should be paid enough to provide for a wife and children—is relatively new. It came about during the late nineteenth century, as production shifted from the home and farm to the factory and office. Women became solely responsible for the unpaid work of child care and no longer contributed equally to the family economy; if they worked, they were paid far less than men.

Anthropologist Helen Fisher describes the stay-at-home house-wife as "more an invention of privileged people in ranked societies than a natural role of the human animal." Yet, although only privileged workers actually earned enough to support wives who didn't "have" to work, the concept took broad root. So did its disastrous corollary: that if men earned enough to support them, women didn't need to make enough to support themselves. The economically dependent wife, purely an upper- and middle-class phenomenon, became the ideal. Marriage became the principal means for a woman to escape the poverty to which low wages otherwise consigned her, and her only passport out of a lower-class status.

During the last decades of the nineteenth century and the beginning of the twentieth century, women moved into the work force in record numbers. The divorce rate escalated sharply too, and social critics were quick to link the two. The census of 1920 reported that over eight million women were employed in 437 categories, from managing offices to hauling freight. Outfitted with jobs and apartments of their own, women did enjoy much more control over their personal and professional lives than ever before. The choice—and it was clearly a choice—between career and marriage became sharper than ever. Encouraged to compete with men while in school, women were nevertheless socialized as future wives and mothers. Women looking for jobs in male-dominated fields were considered traitors to the family, and heroines of movies and popular literature rejected careers for a lifetime of wedded bliss. Even committed professionals, like today's "superwomen," wondered whether they'd be able to do it all, and whether the struggle would be worth it.

A 1936 Gallup Poll asked whether wives should work if their husbands had jobs; 82 percent of the respondents said no (though opposition fell to 63 percent when the husband's employment status was omitted from the question, and even further—to 46 percent of men and 38 percent of women—if there were no children under sixteen). George Gallup chortled that he had "discovered an issue on which voters are about as solidly united as on any subject imaginable—including sin and hay fever." In nearly every state bills were introduced to restrict the employment of married women, and some cities even mounted vigorous campaigns to get them fired. In part these drastic measures reflected the dire economics of the Depression. Popular sentiment held that what work there was should go to husbands and fathers first, but in fact most women found only lower-paying "women's work," and held on to it for economic survival.

As the country headed into World War II, over one thousand state laws denied women access to well-paying jobs. The federal government itself excluded women from 60 percent of all civil service examinations. Personified by "Rosie the Riveter," women in the war effort did make a dent in these discriminatory policies. For the first time, women were welcome in engineering schools, chemical companies, press corps, on Wall Street, and in other traditionally male domains. By the end of the war a married woman over forty was just as likely to be working as a single woman under twenty-five, which was a big change. Many women had to hand over their skilled and well-paying jobs to returning GIs and settle for less prestigious "pink-collar" clerical and service jobs, but they didn't stay home. By 1952 ten million wives held jobs, almost three times as many as a decade earlier, and almost one-quarter were mothers of children under eighteen. In the age group most likely to divorce, 30 percent of women had jobs in 1960; the percentage is now 75 percent.

Over the next several decades, the percentage of women who worked steadily increased; today women constitute 46 percent of the work force. Yet women make only 72 cents for every dollar a man makes—a gain of only 12 cents over the 60 cents to a man's dollar they earned for the preceding hundred years. The wage gap figure is based on full-time workers; it would be even greater if women who are unemployed or who work part-time were figured in. Even working women remain far more economically dependent on marriage than their spouses are. Public policy—the shortage of day care, and a work culture geared to the needs of the traditional male, for whom work unquestioningly comes first—reinforces wives' financial and social dependence. "If men have defaulted on the pact represented by the family wage," points out Barbara Ehrenreich, "so too have their corporate employers."

Traditional marriage serves its intended economic function—as a "mechanism of redistribution," as economist Heidi Hartmann dispassionately puts it—when the breadwinner's adequate wages are fairly distributed to family members. But when these ties are broken by divorce, the buck tends to stop in ex-husbands' pockets. The influx of women into the work force was supposed to change everything, but economics and attitudes lag far behind. Instead, working women face resistance and discrimination in many important arenas.

The Hurdles Working Women Still Face

Myths About Working Women

The image of the woman in the work force as either lonely spinster or pretty gal marking time till she lands a man mercifully went out of style with the advent of Technicolor. But although women on the job have more and more diverse role models than ever, and yes, we have come a long way, baby, damaging myths about women who value their careers still persist.

"WORKING WOMEN ARE UNFEMININE"

It's manly to go out and bring home the bacon, but while working reinforces gender roles for men, for women the opposite is true. Unless they're employed in "appropriate" fields (such as teaching, health care, or fashion), working makes women appear less feminine in the minds of most Americans, not more so. Witness the vilification of Hillary Clinton, for example, who has to keep bringing up her baking skills and devotion to the American family to prove her priorities aren't completely out of whack. Catalyst, a New York–based research company, interviewed 461 high-ranking women executives about their professional achievements. Sixty-one percent said developing a personal style that doesn't threaten male executives was the second most important factor in their success. (The first was to exceed performance expectations.) In the words of one respondent: "Don't be attractive. Don't be too smart. Don't be assertive. Pretend you're not a woman. Don't be single. Don't be a mom. Don't be a divorcée." Needless to say, that rules out most women on the planet.

A working woman faces a no-win double bind: if she plays to win in the work world, she's branded a ball-busting bitch and left high and dry on the romantic front. In fact, notes Susan Brownmiller, "whimsy, unpredictability, and patterns of thinking and behavior that are dominated by emotion . . . are thought to be feminine precisely because they lie outside the established route to success." Saying that she hasn't raised her voice on the job in seventeen years, television producer Linda Bloodworth-Thomason declares, "The minute you ask for a larger filing cabinet you're Leona Helmsley." The qualities that make a working man a perfectionist make a working woman a shrewish control freak, as Barbra Streisand and Roseanne are frequently portrayed.

One reason so many wives hand over the financial reins to their husbands is that it is a way to appear "feminine" (i.e., dependent) in the face of evidence to the contrary (i.e., a paycheck). In most households of women interviewed for this book, men made more than their wives and controlled the family finances. Even when they were generous, this arrangement was a prime source of dissatisfaction for their wives. Marella, who grew up in Trinidad, married a Nigerian man who provided adequately for his wife and three children but refused to let her handle a penny; he did their shopping, he bought their clothes. Marella recalls a single incident that removed any temptation to give the marriage a second chance. Faced with the reality of her departure, her husband finally offered to hand over some cash—but, as always, turned to count it behind his back. Marella said to herself, "Gee, this is part of this man, and he's not going to change," and moved out.

The single-income family is becoming an anachronism, but habits of thought—that men make and look after the money while women stay home and don't mess with math—persist.

"WORKING WOMEN ARE HIGHER RISK THAN MEN"

Another hurdle women must overcome is the family-wage mentality. Men are seen as better employment risks because they have families to support; although wives and mothers do too, they are perceived as less legitimate workers who deserve only "supplementary" wages. Employers fear that women are more likely to stay home with sick children and to stop working if they become pregnant or if husbands get transferred. This does happen, but, conscious of these constraints, women workers tend to be more conscientious and responsible. The true bias is anti-family, not anti-mother; fathers who put their families first, or who stay home because their wives have higher-priority work commitments, are likely to be seen by their employers as insufficiently ambitious and undeserving of promotion.

Biology too is invoked to the disadvantage of the working woman. A *USA Today* front-page story describes "middle-aged career women" who are complaining about memory loss, energy lapses, or lack of focus. "Some—particularly those in high-visibility business positions—are rethinking their careers for fear of *spiraling into incompetence* [italics added]." Yikes—grab those portfolios, boys, before they get left in the ladies' room! The culprit? Menopause, which, the article goes on to point out, affects the functioning of only one woman in ten and which is a completely transitional "ailment." Nor are the fertile exempt from exaggerations about their work-readiness: if menopause isn't derailing her, PMS is making her a bitch-on-wheels.

"Working Women Are Selfish"

Some time after my first child was born, my mother declared with evident pleasure, "You really are a good mother—and I always thought you'd be too selfish." This was a big compliment. Only with the help of therapy was I able to see what it voiced: the resentment and frustration of a woman with an Ivy League degree and a distinguished service record in naval intelligence during World War II, whose considerable abilities had been channeled into child-rearing and a series of interesting but non-paying jobs once her fourth child was in kindergarten. She basically loathed the domestic arena, but had forfeited any other. I chose otherwise, working between naps and feedings, stashing folders in the diaper bag.

Despite the handicaps they face in the workplace, the overwhelming majority of women *want* to work and *like* the economic and personal power that paid employment brings. This fact is still willfully ignored all around. Ambition in women is still not culturally sanctioned, and women who put their careers first are more likely to be criticized than admired.

Women Still Cover the Home Front

Women who stick to their professional guns are expected to throw down their weapons on the domestic front, and they usually do. In her 1989 study of the lives of working mothers, Arlie Hochschild discovered that women worked roughly fifteen hours longer each week than men, the "second shift" to which the title of her book refers. Over a year, this amounts to *an extra month of twenty-four-hour days a year.* Hochschild calls this "an indirect way in which the woman pays *at home* for *economic* discrimination *outside* the home." Accommodating the unpaid demands of the household handicaps women in the job market and exhausts them physically and emotionally. Yet fifty years after women flooded the American work force, the nagging feeling persists that their "real" job is in the home.

A study by Sarah Fenstermaker Berk found that although women did far more of the household labor, 94 percent of the husbands were satisfied with the situation. No news there; what is surprising is the fact that so were 70 percent of the wives, whether or not they worked outside the home. This isn't because a woman is anatomically designed for vacuuming, or better at it, or has more time for it. It's because it supports the image of woman as caregiver, where so much of her identity is rooted, and is reinforced by rhetoric that glorifies homemaking. After all, "the wife who keeps her household running and her husband and children nutritionally and emotionally well-fed no matter what else she does is validated *as a woman*; the husband who is a good economic provider but who never does any housework or child care is still validated *as a man*," observes sociologist Judith Lorber. Women's assumption of these duties, she says, "crystallizes their lesser bargaining power within the household."

Employed or not, nine out of ten of the quarter of a million women surveyed by the Department of Labor in 1995 also said that caring for families was the wife's responsibility. Depressingly, the conclusion drawn by Arlene A. Johnson, vice president of the nonprofit research group that co-sponsored the study, is that this is good news. "This study shows that women challenge the zero-sum model, the idea that if women give more to their work, they give less at home," she says, eager to reassure men that they will still be coming home to a hot dinner and clean kids with no extra effort on their part. Dads who do care for their children compare themselves to all men, not to "working fathers," and tend to give themselves darn good marks. When they care for the kids they get credit for baby-sitting; women get credit only if the children aren't their own. No wonder working women with young children at home, even those with husbands to call upon, are the most stressed-out segment of the population.

Husbands may say they like coming home to a woman who has a paycheck and a life of her own, but almost all still expect her to handle the homemaking. Along with her grueling job as an air traffic controller, Rosemary continued to handle 90 percent of the cooking and child care and housecleaning and bill paying. Once when her husband, Gil, had been unemployed for six months and done nothing but mope on the couch, she spoke up. "I said, 'You have to either take over some of the household stuff or you have to get a job. I can't do it all.' He said, "Oh I knew you'd bring *that* up,' but he did see the reality, and he started taking over the bill paying," she recalls. Not that a couple of hours of monthly paper-work made much of a dent in the domestic workload, and, says Rosemary with a laugh, "That was in 1986, and I want to tell you I am *still* finding things misfiled." Rosemary also worked a lot of overtime, and while Gil didn't like it, he didn't have any problem spending the money either. "So he was real conflicted," observes Rosemary, "and he was real primitive about it, I thought. It was like he wanted me to be a stay-at-home-twenty-four-hours-a-day June Cleaver but he also really liked the income."

Men do more around the house than they used to, but it's sel-dom without a lot of wifely nudging. And while the amount of time fathers spent with their children increased from the mid-1970s to the early 1980s , it averaged out to around one-third of the hours mothers put in. Around the home men tend to do high-profile jobs than can be undertaken when time permits (mowing the lawn, bar-becuing on Saturday night, taking the kids to the zoo), while rou-tine drudgery is left to wives. Women who work outside the home may find that hiring a housekeeper keeps the bathrooms clean, but the overall responsibility remains hers.

Hiring help was Esther's solution when she went back to work in the mid-1960s when the youngest of her three children was eight. It worked, until Sol decided that it cost too much and that he could help her with the housework. "I said, 'Fine, but I don't want to be the one to tell you that it has to be done.' I think that lasted a week. I went back to hiring someone, because [the housekeeper's pay] came off the top of my salary." I had a virtually identical exchange with my husband when we moved into a larger apart-ment. Because I was too pregnant to bend over, and after several weeks of pointing out that we had better things to squabble about than whether he was going to get around to the front bathroom, he agreed that I could hire someone to come in and clean. I was responsible for finding and paying the cleaner.

Both Esther and I assumed the responsibility, logistical and financial, of making sure the house was looked after, and yet our husbands still questioned the outlay. Male or female, those who can afford it buy their way out of domestic drudgery. And unless the housekeeper the wife hires to tidy up after her family is male, she has doubly reinforced the assumption that this is women's work. The same mathematics apply to child care, and to every couple in which the woman laments that her paycheck barely covers the cost of baby-sitting. Her husband is equally responsible for the existence of the baby; why isn't half the expense mentally deducted from *his* paycheck? What does that kind of math prefigure if the marriage ends?

Once on her own, a woman does the same amount of house-work. She still looks after her aging parents and co-chairs the PTA. But with no illusion that these responsibilities are going to be shared equally, resentment and bitterness diminish, which frees up energy and resources for the tasks at hand.

The Double Bind of the Working Mother

At home and at work, the performance of a working mother is measured against the full-timers—men who don't have to make grocery lists during sales meetings, women who have all day to make Play-Doh from scratch. She may be her own harshest critic, her near-heroic juggling act a private struggle with few public champions. Choosing between the demands of the job and those of motherhood is an ongoing conflict that cuts like a knife through the daily lives of working mothers, readily reinforced by internal-ized guilt. Another painful double bind lies in the charge that a woman's commitment to career comes at the expense of her chil-dren, and vice versa.

The either/or construct isn't valid. Of course there is a price, in promotions declined, in school plays missed, in stress between spouses too exhausted to do much besides snore in their rare moments alone. So what else is new? Men are expected to get as much as they can out of life, and ultra-successful males never seem to agonize in public over the fact that adult life is complicated and full of compromises. But when women shoot for the same hoop, the media focuses more on their misgivings than on their achieve-ments, and women give them plenty of fodder. Working mothers may be an ambivalent bunch, but that means they're thinking, not that they're pitiful or morally deficient.

One reason for this ambivalence is an uneasy, class-based distinction, which presumes that poor and working-class mothers work only because they have to (as indeed they do), while higher-income mothers work only because they *want* to, to gratify their own needs or to support weekend houses. As Kathleen Hall Jamieson puts it, "The low-wage mother . . . is assumed to be working for her children; the high-wage mother is presumed to be working for herself." The same act that renders the low-income mom virtuous—the purchase of child care—renders the high-income mom negligent. The low-income woman escapes criticism because she is powerless; she has no options. The career woman, racing home for a little "quality time," shoulders a sackful of guilt and censure. Professional women are covertly routed onto the "mommy track"; except in such "caring" fields as health, education, and social programs, women in upper management are scarce as hen's teeth.

Part-time work is not as well respected as full-time work. The fact that many women work part-time or out of their homes may reinforce the impression that women aren't serious about their careers, but most do so largely to accommodate the unpaid demands of their households. They forfeit good pay, promotions, insurance, sick leave, and paid vacations, and they are more stressed than women who work full-time. For many mothers of young children, such compromises may be necessary and unavoidable. Their resourcefulness and determination in patching it all together is remarkable.

Single mother Annie Lamott wrote *Operating Instructions,* described by the *New York Times* as "the most eccentric baby manual ever." She describes struggling to meet a deadline with her calamine-and-chicken pox–covered baby squirming in her lap. When a friend walked in, she recalls:

> My first feeling was total shame and humiliation that this was such a fly-by-night operation. This is how mothers work. We don't go to our offices and have someone else take care of sick babies. You know those old nudie pens when you tip them upside down and the swimsuit floods? I was flooded with a swimsuit of shame. Just then Sam

woke up crying, and I handed him to my friend and said, 'I'm going to get this manuscript in the mail.' And I ran out in the rain. It took about two weeks until I realized that if this were somebody else, if it were a guy, I would think it was really heroic.

I missed the swimsuit of shame, but I did master the art of typing with a baby in my lap, and developed awesome powers of concentration in the process. I made enough as a writer to be self-employed and to choose how much child care I wanted from year to year, until my younger child was in kindergarten. It was a great luxury. Hannah had far less room to maneuver. She had home-schooled her four children, aged two to ten, on her isolated Wisconsin farm, and staying home with them despite the divorce was her priority.

> I have to admit I was hoping we could get by on ADC [Aid to Dependent Children]. That worked for a while but then they got this thing called Work First, where everybody who's on assistance has to go to work. I had to go to a training program which was very demeaning—you felt like you were in some remedial program for special education or something. Then you were required to get a job. I put an ad in the paper and made up cards for cleaning people's houses, because I thought, "Well, at least I can make my own hours, try to work things out with the kids as best I can, make more money than going out and getting some minimum-wage job." I did that for a quick minute, and it was just too hard for the kids. I didn't have any baby-sitter respond to my ad, so I went in and said, "You know what? This is just pushing me and my kids over the edge. If you're going to cut my assistance"—that was the threat, they'd reduce it and then six months later you'd lose it entirely—"I don't care. Cut it. I'm going to stay at home with my kids right now."

Hannah's assistance did get reduced, but by then, a year and a half after the divorce, both Hannah and her kids were better able to cope. She got work typing reports for a local doctor, whom she moved in with a year later when the house she'd been living in was sold, and subsequently married.

Other working mothers had different solutions, some downright heroic. They switched to night shifts so they could get their kids off to school and be there when they got home. Some took jobs to which they could bring their children. Some were maniacally efficient during working hours so they could have after-school time and evenings free. One self-employed woman raised her rates, got more jobs, and subcontracted them out so she could cut back on her hours. Others, like Phoebe, who is now the administrator of a women's clinic in Bowling Green, went back for their college degrees and relied on family to baby-sit. "I can't count the days when I drove to work on fumes. It was some mean, lean years," she recalls. "There were days when I would give my son a package lunch, and go without myself."

Creativity, stamina, willpower, and luck all play a part in determining the economic effects of divorce, as do women's prior employment and financial experience. But while some pull in six figures and others splurge on ground round, the link between work and positive self-image weaves through each of their stories.

The Value of Work

Work is the cornerstone of identity and mobility in the world at large, and women value it for the same reasons men do. Feeling good about their work makes them feel good about themselves. Married or not, women who work outside the home have higher self-esteem, enjoy more prestige, and are much less depressed than those who do not. Violet, whose husband told the neighbors she was crazy when they remarked on her bruises and black eyes, turned the corner economically and emotionally when she got a state child-care position. "Then everybody knew I wasn't ready for the psycho ward," she declares defiantly. Freud nails it in *Civilization and Its Discontents*: "No other technique for the conduct of life attaches the individual so firmly to reality as laying emphasis on work, for his work at least gives him a secure place in a portion of reality, in the human community." In addition, the income of working wives helps the balance of power in marriages and earns the respect, however grudging, of their husbands.

This is not just true of men, or even of Americans. "Assumptions that women work for just extra cash, or that having a family means giving up a career just don't hold for today's women in

Europe," said a spokesperson for the European Women's Lobby, which surveyed 6,781 women and men in 1995. Six out of ten women surveyed chose to work for money—and three out of five take home at least half of their family's income—but 48 percent said they would still work if money were not an issue. Though their employment circumstances often leave much to be desired, women prefer them to a more leisured but powerless alternative. The major source of stress for working women is not the demands of employment itself, but sexual harassment, followed by efforts to "mommy track" mothers out of their positions. Work benefits families too; dual-income families are more stable financially and maritally.

Work, Identity, and Threatened Mates

Strange things happen to the relation between earning power and personal power when a man and a woman marry. In their study of thousands of diverse American couples, social scientists Philip Blumstein and Pepper Schwartz noted that when men and women marry "it is up to the man to make the *couple's* mark in the world," because society still tends to categorize a woman according to her husband's status. Unlike gay men, who feel more secure with a high-earning partner, a partner's financial success is no comfort to a heterosexual man; just the opposite is often the case. Unlike lesbians, who believe both partners should be financially self-sufficient and encourage each other in that direction, heterosexual men compete with successful partners. "This makes us think it is not women's working and achieving that causes problems between partners. Rather, the source is the difficulty men have in accepting female equality," conclude Blumstein and Schwartz.

A man's right to self-sufficiency, financial and personal, is unquestioned. So, now, is a single woman's. But the prospect of a married woman exercising that same right—to spend money as she chooses—is problematic. Few women—even those who made as much or more than their mates—seriously challenged their husbands' control of the finances, with unfortunate consequences. As Arlie Hochschild puts it, "if money is the key organizing principle to the relations between men and women in marriage, it's a pity, for men because it puts their role at home at the mercy of the blind fluctuations of the marketplace, and for women because if money talks at home, it favors men." In granting their husbands control over how money would be spent, wives—consciously or not—also yielded the majority vote in decisions that had little or nothing to do with money. The fact that women made important decisions all day on the job became irrelevant at home.

Husbands may pull the purse strings tight, but women also handicap themselves in their scramble to protect the legendarily fragile male ego. Doing the housework, maintaining that they work only to help their families, accepting lesser-paying jobs segregated by sex, are all ways in which they compensate for their economic power—part of the "balancing" behavior described in the first chapter.

Working women may be reluctant to throw their financial weight around, but men customarily equate financial achievement with the right to control how that income is spent. (A major reason child support goes unpaid is that ex-husbands resent not knowing just where the money goes.) Spending decisions are, of course, about power. Blumstein and Schwartz bluntly concluded that except among lesbian couples (who use it to *avoid* dependence), money establishes the balance of power in romantic relationships when the partners live together. "He was always crying poor" is a refrain that runs though my interviews, as did the complaint that there often seemed to be enough money to fund the husband's pursuits but not the wife's.

Many divorced women felt better off simply because they were no longer oppressed by their husbands' miserliness. My husband was a veritable Rockefeller compared to Jodie's, who never once picked up the restaurant tab. "He would always say, 'Your portion is X,'" she recalls ruefully. A marriage counselor tells me this is not uncommon, but this guy takes the cake. The night before the movers came, Jodie, the software analyst, heard Paul rummaging through her boxes with a flashlight. He was searching for her scale, to take out the battery he'd paid for. Wearily she told him, "I already *took* the goddamn battery out of the scale." The creepy part, as she admits, is that she'd known he would want it back.

Rachel's husband was more stingy with authority than money; Jonathan was happy to shell out for extras in addition to the $10-a-day "allowance" he provided his wife, whose income as a salesclerk was negligible compared to his salary at an investment bank. His response to her repeated requests for a bank account of her own was "Why? Didn't you just get shoes last week? Do I deny you anything?" He never understood her discontent, nor did he ever grasp why money was such a big issue when they went to counseling. Marella's husband made the same case whenever she asked for spending money of her own. "Why do you need money?" Reg would ask. "I'm bringing food in the house, I'm giving you things." He did buy her whatever she needed, "but *still*," asks Marella rhetorically, her Trinidadian lilt still perceptible, "where was my independence? I felt I was losing it. And if I lost my independence, I would lose my identity."

A balance can be struck when partners genuinely respect each other's contributions to the marital economy. It doesn't matter who does what, and meeting a deadline is no less grueling than getting a rambunctious toddler through the day in one piece. But striking such a balance between what each partner brings to the table requires generosity, confidence, and a willingness to defy social norms, qualities in short supply in strained marriages.

My husband and I couldn't balance our respective contributions to the marriage, or indeed even conceive of the need to do so. When we married we both had full-time jobs, and I figured we always would. Instead, I wrote a little paperback which, much to everyone's surprise, turned into a best-selling series. My husband's urging and support were crucial to the decision to become self-employed, and he and I became partners in the book business. It was a way around the threat my success posed for him and for our marriage. I accepted it unquestioningly, though in the long run it was a mistake that brought out both our weaknesses. My husband was scrupulous about crediting me with the idea for the series. I was equally diligent about telling everyone I couldn't have done it without him, and largely convincing myself in the process. I never acknowledged, even to myself, how much his lack of independent initiative bothered me, but it rankled more with each passing year, as the royalty income from the book series diminished. The low point was our desultory tenth anniversary, when my husband proposed that rather than either of us taking jobs we didn't like, we should live off the interest from our savings. Even if the income had been remotely sufficient, his attitude chilled me: in hard times, he wanted to hunker down and spend less, while I figured we should go out and earn more.

Caroline's husband's lack of ambition bothered her too. A hard-nosed British-born businesswoman, she's candid about the fact that her decision to divorce after eight years of marriage was "very much a material thing." The engineering student she'd married at twenty turned out to be thoroughly disinclined to put his nose to the grindstone. Their apartment was tiny, they were basically broke, and Caroline had bigger plans. She describes her marital situation as "the classic thing: once my daughter was about two and a half, I could see that I could cope on my own. I thought to myself, 'I'd rather be poor on my own than poor with somebody else.'" Out she went, hung her name on the door, and opened her own floral-design business.

A more common cause of stress crops up when traditional earning roles are reversed. Money wasn't an issue in Rosemary's marriage until her steadily increasing air traffic controller's salary overtook her husband Gil's stagnant one, and "although he always denied that it was a problem, it was a problem. I think he felt emasculated by that," she observes. "In arguments, he'd bring it up scathingly. I'd say something like, 'Yeah, I know I'm not home as much as you'd like me to be, but that's because I'm at work,' and he'd say, 'Oh, you'd have to bring up money, wouldn't you?'"

Anneke's husband George maintained the balance of power by a kind of emotional sabotage, systematically demeaning the work she loved. She had been writing in her free time for several years, and when she got laid off, they decided that she should try her hand at it full-time. Anneke's husband hated the romance novels she wrote, dismissing them as "trash," but he figured that maybe she'd produce a best-seller and enable him to quit his boring job as a computer systems analyst. Unfortunately he was both envious and critical, accusing her of sitting on her ass all day and expecting her to run his errands "because," she explains, "what I did was not really considered work." With six books and ten years as a writer now under her belt, Anneke, who lives in Tucson, figures she's probably making about 25 or 30 percent less than before she became self-employed. "I have to watch what I spend a little more than I used to have to," she says with a smile, "but I don't mind. It's not a big price to pay for freedom."

Terry McMillan, the author of *Waiting to Exhale*, gets to the heart of it with the observation that her son's father was unable to see that her achievements diminished him only in his own mind. "When I was writing my novel, he was like, 'Okay, do your thing,' until the book was actually published," recalled McMillan. "Then he had to look at things a little differently, and he didn't want to. I tried for three years to get him to understand that what I am doesn't make you any less of what you are. But he couldn't, and I had to leave him ... " A similar dynamic fueled my husband's unwillingness to live, even temporarily, in a country where only I spoke the language. That kind of reliance was unthinkable to him, and made him feel reduced by comparison. Sadly, these men perceived *inter*dependence, a vital and mutually nourishing element of marriage, as dependence, a threatening dead end.

Work and Independence

Conservatives who blame the decline of family values on women in the work force have it wrong. Work is not the culprit, nor is the relationship between women's employment and the divorce rate as obvious as it appears at first glance. The influx of women into the job market at the end of the nineteenth century did make it possible for the first time for women to survive economically outside marriage, and divorce rates did climb. But simultaneously, as housewifery emerged as a full-time occupation, the proportion of married women in the paid labor force fell. Commenting on this fact, divorce historian Roderick Phillips observes that "it is not always clear whether employed women divorced or divorced women obtained employment." Getting divorced requires an independent income, but earning power does not inherently predispose women to divorce.

In fact, just the opposite is usually the case. Far from causing divorce, notes Hochschild, "in many ways, the fact that wives work both benefits and stabilizes marriage." Not only is a working woman more likely to be happy and healthy, but her income protects the family against the terrible disruption of poverty. Dr. William Julius Wilson, the eminent scholar of race and poverty in America, writes that work "provides the anchor for the spatial and temporal aspects of daily life. It determines where you are going to be and when you are going to be there. In this absence of regular employment, life, including family life, becomes less coherent."

Work *reduces* the likelihood of divorce, which is three to five times more common among couples who live in poverty. The happier employed women are with their jobs, which provide gratification and stimulation they would otherwise be relying solely on their home lives to provide, the happier women are with their marriages.

On the other hand, unlike stay-at-home housewives and mothers, working women are less likely to be hostage to their marriages financially. Though they are likely to exercise it only when unhappy or exploited, they enjoy a degree of control over their circumstances that men have long enjoyed. As Willa, the TV producer, puts it, "Work is my declaration of independence." Asked if anything made their divorces easier, many women cited the fact that they could earn their own living. "I could afford it," says Willa, for example. "I didn't have to eat shit, or really change my lifestyle."

If a single theme asserts itself over the course of these interviews, it's the crucial importance of a skill that brings with it some degree of financial independence. The plight of women who married very young, or who dropped off the employment ladder to tend to children and mates, is all too familiar. That's why Tory's advice to anyone contemplating divorce is to "get a part-time job. Start validating your ability to get out in the world." In the experience of this fifty-two-year-old marketing director from Gaithersburg, Maryland, the happiest marriages are those in which the woman derives a sense of identity and achievement for jobs well done outside the home, paid or unpaid. So when there was an opening in her office for a meeting coordinator, Tory urged her assistant to apply.

> She said, "I'm just a secretary. How would I do that?" I said, "Tell me how you went on vacation with your six children." She said, "Well let's see . . . I canceled the paper, took the dog to the kennel, loaded both station wagons, packed sandwiches where I could reach them, checked the oven, lowered the thermostat—" I said, "Sales meetings'll be a piece of cake." This woman had never valued those things because no one had ever paid her for them, but all those skills were totally transferable.

Tory's assistant did in fact become the meeting planner, and a very good one.

Claire didn't reap the nonmaterial, or even the material benefits of employment outside the home until she was sixty-three, when, recently divorced, she got her first paying job as a bank trustee. "Men treat you differently when you're getting paid," she observed happily. "I like it a lot." Monica recalls a phone call from a friend who got a job in a doctor's office after having been a homemaker all her life. "I couldn't believe someone would really pay me to do this," crowed the friend. "I knew I had value; it was $187 a week. No one ever gave me money to cook dinner."

Tory herself had walked out on her marriage at age eighteen, with no money, no education, and a six-week-old baby under her arm. Looking back on it, Tory figures the only reason she did it was that she was too young to know she wasn't supposed to be able to pull it off. She feels fortunate that she had a vision of what her life should be like, and the natural abilities to make it happen. When her son was small, she remembers, people would say, "Why don't you get a waitress job, or a hat-check girl, or barmaid? You'd make more money than in an office job." Tory says firmly, "my vision was: that's not who I am, I'm a white-collar worker and I want to be in an office." Thirty-three years later, she makes $145,000 a year as the human resources director for a *Fortune* 500 company. Tory feels like she beat the odds, and in a very real sense she did. Her brief marriage made her truly value her independence, and reinforced the courage not to compromise it.

It was Tory's refusal to give up her financial autonomy that became an issue with the man she lived with for ten years—"it was more like a marriage than my marriage"—and wanted to marry. If they had an argument, he used to threaten not to buy Tory dinner, whereupon she would say, "Fine, I can buy my own dinner." The more she made, as Tory came to understand very clearly, "the more power he was losing, and it was very distressing to him. As proud as he was of my success, it was also undermining his ability to sweep me off into another world that I couldn't go to by myself." He agreed to marry her if she quit her well-paying job so they could travel freely. She agreed on one condition:

> I said to him, "If it doesn't work out, ten years later I'm going to have a problem reentering the work force. So I want money in an escrow account, whatever amount we agree upon. It will only turn over to me if we get a divorce, but I'm not going to try and fight you

through a divorce because you're a lot bigger and more powerful than I am. Then you would know every day when I got up that I'm here because I want to be here." He said, "You're treating this like a business agreement. It's terrible, it's so cold-hearted."

He wouldn't agree, she wouldn't give in, and it broke up the relationship.

Wendy went from being a completely dependent doctor's wife to a successful freelance writer and cookbook author. Although she had published several books by the time her marriage finally dissolved, the divorce was liberating personally and professionally. "It was wonderful, because I had the time," she explains. "You know, in a marriage you have to negotiate everything: what you're going to have for dinner, what you're going to watch for television, what time you're going to bed. And when you live alone you live your life exactly the way you want to." Her income has dropped precipitously, but she describes the little townhouse she now lives in as "home in a way that big house never was."

She's written several more books which have sold quite nicely, though freelance writing isn't exactly financially secure. With a wry smile, she admits that "I'd probably still be married to him if I hadn't gotten a profession and a way to make money." Wendy remembers the first time she sold a magazine article, for $400.

I remember being very proud of that money. I was standing in the kitchen and I showed him that check. He said, "Oh that's wonderful, Wendy! You go buy yourself something beautiful with that." I bought a new dishwasher. The old one had broken; the timing was right. It made him mad. I thought that he wanted to relegate that money to something frivolous; he was the wage-earner, he would supply the dishwashers.

She made the same kind of choice in paying half of her son's and daughter's law school tuitions, though not legally obligated to do so. "I really couldn't afford it," she says, "but it was so important to me for my children to know that their mother had done that, even though we never really talked about it." Thirteen years later, having spent all that money to help educate her children is what she is proudest of.

In my case, divorce purchased liberation from the bickering that cropped up endlessly around nonessential spending: whether to take a cab instead of the subway, to rent a car for the weekend, to get air conditioning in the bedrooms. It was absurd for a couple who worked at home through the muggy New York summer not to have air conditioning, as my therapist gently pointed out. "Just tell him you're doing it." she suggested. Eventually I scraped up the courage, though only after soft-pedaling the decision by getting five competitive bids to prove I wasn't being extravagant or impetuous. On those hot days it sure makes me happy, but at the time it seemed a Pyrrhic victory. Those kinds of struggles motivated my insistence on sole control of a bank account into which each of us puts a flat annual sum to cover kids' clothing, after-school activities, and miscellaneous expenses. He was bitterly opposed, but I think several years' evidence that my essentially frugal nature hasn't changed have calmed his fears. Every time I write a check without having to haggle or confer, I realize that digging in my heels on that particular point was worth it. Financial independence buys freedom—and the responsibilities that come with it.

Conquering Fear and Ambivalence

Although it seemed completely irrational, money took on tremendous symbolic importance when I decided to end my marriage. As sensible friends kept reminding me, I had been supporting myself, my husband, and our two children quite well for a decade. This didn't keep me from lying awake at night in a cold sweat, wondering how on earth I'd make the mortgage payments. I wasn't alone in these fears. When Willa and Desmond separated she was a high-profile television producer pulling in a six-figure salary. "Yet I didn't have any confidence, I'm wondering if I can make it on my own," she muses in retrospect. "Someone had to tell me, 'Stupid! You've been managing all along, you big dummy.' I was frightened to death. I got a mortgage on my own, and I was fine." Women often pass through a period of intense conflict on the way to autonomy.

Financial independence for women, a tenet of modern feminism, was a radical idea when it surfaced in the late nineteenth century and remains surprisingly so to this day. It is a two-edged sword, one ably wielded by the forces that defeated the Equal Rights Amendment. "By simply asserting women's right to enter the labor market on an equal footing with men, feminism undercut the dependent housewife's already tenuous 'right' to be supported," writes Ehrenreich. "If some women can 'pull their own weight'—as a resentful husband or a female follower of Phyllis Schlafly might reason—then why shouldn't all of them?" Some women prefer not to work for pay, and resent the diminishing status of the full-time homemaker; they're anxious that if they leave the kitchen, they won't be able to get back in. But women cannot reap the benefits of independence without accepting financial and personal responsibility for themselves, just as men are expected to do. As they said in Maoist China, women hold up half the sky. Equal rights mean equal responsibilities.

Though her health and self-esteem have improved immeasurably, Keiko is one of those who is ambivalent about supporting herself and struggling with the reality that she can't have it both ways. Although she retains many of the trappings of her married life—the big house on Seattle's Lake Washington, the son in private school—they now cost her directly, and Keiko is both fearful and resentful of having to support herself. "I never thought the financial side of things was that important," she declares, "but my son will never be anything as silly as a music teacher." And if his heart is set on it? Then, says Keiko, a striking Japanese-American in her mid-thirties, he'd better be plenty good at it, good enough to make real money. The cardinal lesson of Keiko's divorce is crystal clear: "Achieve financial independence before you're married and maintain that financial independence while you're married." Keiko has vowed not to remarry until she feels financially secure, "because I don't think I could remain composed in a marriage if I ever felt this could happen again." She adds, "I stayed in that marriage much longer than I should have because I was terrified of being without an income. And I still am."

Married to a lawyer, Keiko has seen her income drop precipitously and is far from sanguine about it. It's been two years since her divorce, and every so often she thinks about giving in to her ex's persistent suggestions that they give it another try. Paradoxically, it's the money issue that makes her say no, "because if he had no compunctions about leaving me in these financial straits, never again could I see him as a compassionate person, as a person that ever valued me, because you don't do that to someone that you love," she declares angrily. "In a sense you have a better sense of who that person is when you divorce him than when you're married to him."

Keiko's husband, Rick, made good money, and when she realized she wanted to leave the marriage, she switched from teaching oboe and clarinet to carve out a more lucrative career in real estate. Keiko gets almost $4,000 a month in child support and alimony. It's not adequate, she says, "not considering what I lived on." What really rankles is that the alimony payments diminish and then end after four years, "and he *knew* I'd have to change professions," she says bitterly. Afflicted by depression, insomnia, and severe psoriasis during the last years of her marriage, Keiko loses sight of the fact that her former way of life cost her her health and her sovereignty.

"Women lack a positive emotional vocabulary about money," notes feminist theorist Naomi Wolf, who feels that not until the 1980s did women begin to investigate "the expanded sense of personal importance, choice, mastery, and psychic empire building that money symbolizes." Unlike men, who grow up knowing they're supposed to sally forth to seek their fortunes, women are more passive and conflicted about self-sufficiency. Even when a husband's financial contribution is negligible, his emotional and symbolic shelter is not. In my marriage, I made the money but had a hard time believing it, let alone taking credit for it. But as I faced the realities of going it on my own, I could no longer afford to be coy or ambivalent about my earning power. More importantly, I no longer felt the need.

For Keiko too, the need to support herself was coupled with real emotional growth. Resentful though she is, Keiko nevertheless cherishes how employment makes her feel. "I think that work is very important for me, it's really sort of everything. I think I am stronger now because I have to work; I didn't really have to work before. I'm taking responsibility for my needs in a way that I've never had to in the past," she notes perceptively. "I'd go into interviews and say, 'I can do this work, I can learn this quickly, I can get these licenses fast.' And I *got* all the licenses fast, and I got the listings." Keiko's new job is hard work, and it's a constant scramble to get new clients and to weather lulls in the market, but her career in real estate is progressing nicely. Where did the confidence came from? "I just felt so much better about myself," she explains. "I felt more capable, I felt as though I had done what I should have done years ago, and I felt that once I had taken the risk of leaving the marriage that I could take the risk of doing something radically different in the way of work." Risk engenders risk, and the payoff is all Keiko's.

When a woman faces actually going it alone, a link between "unmarried" and "destitute" keeps surfacing like an annoying car alarm. It's probably not being stolen and it's certainly not your car, but it keeps waking you up. Some never learn to sleep through it, but it gets easier as time goes by and the evidence of survival and even prosperity accumulates. No one and nothing can take that away, or the genuine self-esteem that comes with it.

Lifestyle Changes After Divorce

Living Less Well—"We'd have a fried egg and we'd split it three ways."

As noted earlier, divorce frequently results in a lower standard of living, especially in the short term and especially for the mothers of young children. This holds true for professional women as well as poor ones. Across the socioeconomic spectrum, women had to tighten their belts and scramble to make ends meet, especially in the years immediately following their divorces. Yet across the board reduced circumstances translated into a better quality of life, because now women were in control of those circumstances.

Sophia's husband forbade her to work or drive, kept track of her whereabouts, and, like Hannah's husband, made her submit the grocery receipt—the only money she was allowed to handle—for his approval. When her marriage ended, she refused to go on welfare and got a job as an information operator with Pacific Telephone. Sophia's answer to whether she and her two daughters lived as well after her separation is "Oh God no. We'd have a fried egg and we'd split it three ways." Nevertheless, that "didn't bother me," Sophia maintains, "because I was happy, and my girls were happy. I always had money for rent, for utilities—we were very frugal, we watched everything—and my daughters weren't demanding, even when they were little. It was like there was a breath of fresh air: they could be themselves, they could play, laugh, go for walks." Twenty-eight at the time, Sophia met her second husband at a union meeting and remarried two years later. In 1986, after eighteen years of marriage, she opened her own travel agency, and now drives a silver Cadillac with vanity plates.

Marella had worked in a bank before getting married, and relished the thought of becoming financially autonomous again. The job she found working in and subsequently managing a video rental store made many precious things possible: "to have my own credit card, to say what I feel, to bring out the best in me without anyone looking over my shoulder or not giving me encouragement." Her husband, Reg, expected his wife to stay home with their three children. "When I would say, 'Gee, I need to go out and find a job,' he would say, 'How are you going to get a job? You don't have any skills.' These statements *were* true, but where was the support?" recalls Marella, with a sideways look and a tilt of her strong jaw. "I said, 'To heck with this.'" She was able to open up her own franchise five years later, but it went bust when the neighborhood declined. Since then her finances have had their ups and downs, but she's always made ends meet and had a little something left over for birthdays and holidays.

Reg paid no child support, and Marella could never have made it without the help of other members of her family, who had emigrated from Trinidad in her footsteps. (She had twins out of wedlock two years after leaving her Nigerian husband.) Timely gifts of sneakers, shirts, and underwear made all the difference, "and there were many times when I stayed without in order for them to have. I mean, literally walking the streets with nothing in my pocket. Sometimes I felt really, really badly about it," she admits. "I felt like, 'Am I slipping? Am I going to make it? Am I going to lose it?'" It was frightening, but Marella's faith and native optimism would weigh in. "I've always managed to bounce back, on my own," she says, "and I would pray a lot."

A low point with an absurdist twist was Marella's run-in with the IRS. "I had a tax audit, can you imagine?" she recalls with a hoot of laughter. "The IRS called me, someone said, 'This is impossible.' She said, 'Either you are lying or you are doing something illegal. There's no way you can do all these things on your income.'

But as I said, my family really supported me." Another absurdist low point was Marella's encounter with the credit system. "I had nothing in my name and it was very frustrating. I kept applying and they would reject me. I called up [the credit agency] and said, 'Why are you denying me credit? What's wrong? I don't owe anyone.' She said to me, 'That's the problem.'" So Marella got a bank loan, not because she needed the money but purely to establish credit.

Her third child, a babe in arms when Marella walked out of her marriage, is now a sophomore at a top-flight New England prep school. "You think I can afford that? She got a scholarship. The struggle was worth it. And no one knew that I didn't eat any lunch," says Marella. She's not boasting, just telling it as she sees it, even though she's currently on unemployment and job-hunting hard.

Marella's one regret is not having pursued a graduate degree while she was still married. "I think my life would have been different if I'd gone to school then, because I really wanted to go into teaching. I wouldn't have been unemployed right now," she says with a laugh, admitting that she went to get a transcript on her way to that morning's job interview. Beverley, who at twenty married a dapper young minister she met in church, agrees. She thinks she'd have been able to finish school had she not divorced, and "*that* affects my employment status." Now an insurance claims processor, she says, "The effect of the divorce on my lifestyle was tremendous. I still have moments when I get angry at him again, but I'm better about it now. We've been divorced since 1990 and that's just the way it is." She and her son just moved out of the one-bedroom apartment that was all she could afford following her divorce, and money's still tight. She used to blame it on Jermaine's deadbeat dad, but now, she admits, "I am getting support, and I still don't see the income level I'd like to. But at the same time," she adds philosophically, "we make choices about how we live and what we want."

Remarriage boosted a number of women up the financial ladder, but many took years to regain their previous standard of living. Some have had to settle for a reduced lifestyle unless, like Keiko, they were able to switch to more lucrative careers. Gloria, a medical technician, makes a quarter of what she and her periodontist husband lived on. "I had a housekeeper, and a gardener. Now I'm buying Lady Lee Raisin Bran," she says tartly. So does she wish she'd stayed married? "Fuck, *no*," is the unequivocal reply. "It was a fancy place where I could have everything I wanted and all the clothes I wanted and go on the trips and be a fancy doctor's wife, but I was squelched inside. I was ironed flat." After five moves in a year, Gloria recently bought a small house, where, she declares, "I feel like queen of the castle. I'm very, very proud, because I've done it with my own two hands, and I never did that before."

The fact that divorced women happily adjusted to reduced circumstances doesn't justify the economic discrimination and enforcement loopholes that so greatly contribute to it. But the sense of pride and achievement that accompanies financial self-sufficiency is beyond value, which is why the mathematics of divorce often defies rational analysis.

Living as Well or Better—"I didn't have to pay for him, and he had very expensive tastes."

Quite a few women pay all the bills, and get divorced partly to rid themselves of their husband's dead weight or fiscal irresponsibility. For those, the financial picture may clear and even brighten after divorce. Wives often bring home half the bacon, and some the whole pig. More than half of the 1,502 working women surveyed in a 1994 poll—including 48 percent of those who were married— said they provided at least half of their household's income. Colleen Keast, executive director of the Whirlpool Foundation, a sponsor of the study, declared, "This calls for an end to the debate over whether women should or shouldn't work"—as though it were a matter of choice, and as though the decision were being made by some higher authority!

This is corroborated by the Bureau of Labor Statistics, which reported that married women who worked full-time through 1993 contributed a median of 41 percent of the family's income, and earned more than their husbands in 23 percent of the families where both spouses worked. Nor is this an American phenomenon. Three out of five working women surveyed in Europe in 1995 take home at least half of their family's income.

Francine, who married the Danish graduate student, is an exam-

ple of those childless women with full-time jobs for whom divorce
causes no financial hardship. Nevertheless, there is usually a tem-
porary period of adjustment to shouldering basic living expenses
that had once been shared, and of reestablishing credit. Now that
her single salary was covering the rent, Francine remembers thinking, "'Gee, I wish I could stick out [the marriage] for another year,
and be financially liquid,' because I was paying off credit card
debt." But did her standard of living change significantly? "No, it
really didn't." In fact, as the primary breadwinner she was grateful
not to be paying Frederik alimony.

A growing practice as a massage therapist and a small inheritance at the time of the divorce brought Tara's income up to the
level of her photographer husband's, so her standard of living
changed very little, but she ran into a common consequence of
divorce: the loss of credit. Toward the end of her marriage she and
Hal got a Visa card. Because she made a big deal about wanting her
own credit rating, Citibank assured them that both could be primary cardholders, but "when we divorced, Hal got the credit line."
Then the phone company informed Tara that they would have to
do an entirely new credit rating on her, "even though the telephone
had been in my name for the entire marriage, and, as the woman
said, 'And I bet you paid the bills, right?' I said, 'Yes, I did,'" recalls
Tara.

Susannah, the preppie film archivist, finessed the credit problem
twenty years ago by simply applying for a card under her maiden
name before her divorce had gone through. "Then I had it, so what
could they do?" she points out, grinning. Her standard of living
generally improved. First of all she wasn't paying child support for
his kids from his first marriage anymore, and secondly, as she
explained, "I didn't have to pay for him, and he had very expensive
tastes. If he needed a new pair of glasses, he couldn't just go to
Lenscrafters."

Laurie, the homemaker in Reeboks, recalls with chagrin that her
husband took her name off their Sears card without even telling her.
She went on paying it until a humiliating incident set her straight—
she and her son got up to the register with his entire back-to-school
wardrobe and the sales clerk informed her the card was no good.
Taking matters into her own hands, Laurie dealt with losing credit
with aplomb. She took a clerical job, worked two shifts as a wait-

ress, bought herself a book called *How to Get Credit for Almost Anything,* and "followed it like a religion."

> It worked. I got credit. I bought an Electrolux vacuum cleaner on credit, then I went to Macy's and got my own card. Then I was driving to the supermarket one day and I realized that if I got to the market and parked it I'd never make it home on time. (The car had a little electrical glitch: if you turned it off, it had to sit at least two hours. Of course I had my cables, and everyone in the neighborhood knew me because I was always standing on some corner waving my cables.) The supermarket was on the way to a Datsun dealership. I drove past the market and my son said, "Where are you going?" and I said, "I'm going to go buy a new car." He said, "You can't do that, you don't have any money." I said, "Watch me. I'm gonna do it." And I did!

Still, things were tight until Laurie remarried three years later.

Getting credit under her own name wasn't a problem for Lorraine, but half the family income went out the door with her husband Jim. But most of his $30,000 earnings had gone into one ill-fated business venture after another while Lorraine's salary as a second grade teacher essentially supported the family. Jim does pay child support, $125 a week for their two boys. In the wake of the divorce, Lorraine goes out to dinner less often—his supplementary income did cover some small luxuries—but feels that her situation hasn't really changed much. As she puts it, "Being able to choose where the money goes makes having less money work better."

The Career Benefits of Divorce

Most of the women interviewed for this book would not define themselves as "career women." Most would have put the fact that they were married higher on a list of defining characteristics than the nature of the work they did. For some, work might never have become a priority if they hadn't had to go it on their own. But regardless of the degree of choice involved, many careers were developed, changed, and jump-started. Having had the courage to change their lives bolstered the confidence of these women in themselves and in their ability to deal with the world at large. The great majority, having already jettisoned an enormous amount of personal baggage, discovered that traveling light made taking professional risks much less farfetched.

That was certainly the case for Sharon, whose tawny skin and thick black hair bespeak her Hispanic heritage; her good looks are all the more striking because of her obvious diffidence about them. "When I was married I felt that every professional decision had to be figured into the formula of the marriage. That's not a bad thing, but it felt absolutely fantastic to be free of that constraint. I felt that I had no limits in my work," she declares. Now thirty-five, Sharon happened into a job with a commercial bank and soon discovered a real talent for pension fund management. After she and Charles separated, Sharon's income, which had been steadily increasing all along, went up to six figures. "The divorce helped in making me realize that we are pretty much responsible for our own destinies," she says. "I love working. I hope to work till I'm eighty."

"Marriage and divorce put me in a position of really having to become seriously independent," says Susannah. "The career wasn't just an afterthought, it was really something that I had to do." Born to privilege, Susannah figures she might well have ended up with a Volvo wagon and a part-time job in the local art museum had her early marriage to a much older composer not derailed that suburban scenario. Now a well-respected film archivist and confirmed city dweller, she is intensely grateful for the way things worked out.

Similarly, Helen claims that "the divorce affected my lifestyle because it made me get my act in gear!" Until his drinking took over, her husband, Ron, made more than she did, even when they were working on the same assembly line. "I know that because at one point we had the same boss, and I got paid less than he did and I was actually doing more skilled work," she remembers. But with only a high school education, when they moved to Florida all she found was part-time work at the minimum wage. "I didn't have any marketable skills other than what I could do with my back or my hands," she explains. Helen remembers feeling afraid. "I was the kitchen manager of a state-funded preschool in Woodville, outside Tallahassee. I worked twenty hours a week, and I thought, "If I'm going to be out on my own, I need more." Helen would have been in real straits if she hadn't been able to rely on her parents, who let her move home, "but on the condition that I get back on my feet and get out there and start living again. And that's what I did."

> I got a job, and I loved it, working as a part-time cashier in a hardware store. I had to get a roommate when I moved out, which was fine at the time. But I had set a long-term goal: to really prove to myself that I could take care of myself, be self-sufficient after all these changes I had made, and that meant having my own apartment. When they opened up a new store I got a job as a full-time cashier—I had to relocate into Silicon Valley, and I did—and inside of two months I got promoted to security cashier, balancing all the trays.

Promotion in hand, Helen went out and rented her own place.

As noted, an impressive number of women completed their educations after divorce and went on to get meaningful jobs they really liked. A 1991 study in *Working Woman* magazine confirms that divorce invigorates many careers. "Career women participating in the study tended to boost their output, win higher performance ratings, and feel more motivated and satisfied with their jobs after a marital split," reports Andrea Warfield, associate professor of management at Ferris State University, who conducted the study. Why the difference? According to Warfield, "Women still tend to be the custodians of relationships, which takes a lot of time and effort. So when a marriage dissolves, there is a tremendous amount of energy freed up that can be redirected into a career." Whether the constraints are externally or internally imposed, unguessed-at abilities burst out when these constraints no longer hold women back. Whether galvanized by fear, ambition, or orthodontist's bills, these women are forging ahead and doing very well for themselves.

Why Divorce Pays Off

The bottom line? Even women whose standards of living declined significantly prefer their present situation to the economics of an unhappy marriage. This is corroborated in Dr. Constance Ahrons's study of 379 divorced couples, in which she writes, "Much to my surprise, many of these women said they felt more satisfied with their financial situation after the divorce." It seems it is better to be solvent in rented rooms than powerless in a palace, especially when that rented room is only a springboard to a better situation.

Anthropologist Helen Fisher notes that divorce is common in societies where both women and men own valued goods and have the right to distribute them, whether the riches take the form of stock certificates or camels. "Powerful working women will almost certainly sustain the long-term trends initiated by the Industrial Revolution: later marriage, fewer children, more divorce, and more remarriage," she predicts.

The best partnerships are those in which each person brings something of equal value to the table and hangs in there out of his or her own free will. One or both of those conditions were missing from these unhappy marriages, which the glue of financial dependence was no longer strong enough to hold together. Because public attitudes have yet to catch up with social and economic reality, couples need to figure out their own scenarios in which interdependence replaces dependence, money talks but doesn't yell, and respect for each other's contributions is authentic. The challenge is strewn with cultural and personal land mines, and it's also cause for celebration.

CHAPTER 5

Spirit
and
Sense
of Self

A woman contemplating divorce wonders who she will be on the other side of the marital divide. How long will the transition take, and at the end will she like the person she has become? Though she can learn from the experiences of others, ultimately each person must take the plunge, finding out for herself what bumps and lumps and giddy pleasures await her along the way. For Lorraine, the path of emotions started with "fear of 'Is it all going to work out?'" Then, "I felt guilty for messing up his life." And afterward? "Yahoo! Sorry, I can't help it. A sense of liberation."

Asked "What was your primary emotion when your marriage ended?" the women interviewed replied "fear" and "relief" in nearly equal numbers. The relief was permanent; the fear was not.

Over time women passed through panic, grief, and anger to happiness and calm. The only men who still believe that female veterans of divorce are somehow damaged goods are the ones in the market for a malleable virgin bride. Women who have divorced are indeed older and wiser, but they are not sadder. They possess an unshakable sense of themselves, and they are enjoying the pride and confidence that accompany it.

Wifehood and the Loss of Identity

"I'll tell you what went wrong with your marriage," says my frank friend Dietshe. "You let your identity slip and then you got mad." In retrospect my ears should have pricked up when the minister proclaimed my husband and me "man and wife." The groom's biological and social status stayed the same, but I had mutated from woman to wife. Though I was fortunate to have and return my husband's love, the shift was ultimately not to my advantage. For me as for so many others, it meant reconfiguring myself into a less potent stranger, and struggling with an inarticulate discrepancy between what I felt and how I was supposed to feel. As discussed in detail in Chapter One, many women "dwindle" during marriage, turning into anxious, muted versions of their former selves.

A sense of self seems an odd thing to mislay, but even for those fortunate enough to have acquired one relatively early on in life, marriage can erode it. When *New Woman* polled four thousand married readers in 1995, editor Stephanie von Hirschberg reported that "more than half of women in bad marriages mourn a lost sense of self." Sadly and inexorably, being wives and being themselves turned out to be mutually exclusive. Von Hirschberg went on to say, "When asked to pick the worst thing about their marriages, a large proportion of these women said 'not being who I really am.' Not surprisingly, they are more likely to say they 'play the role of Perfect Wife' than women in happy marriages." Role-playing works on stage, but in real life it grows hollow and exhausting. It may offer protection, but in the long run its shelter is illusory.

Women Who Have Ended a Marriage Learn Who They Are

TARA

When you separate, really think about what your motives are, and how much you're projecting onto your husband. Who will you be when you're not married anymore? When you're alone, who will you be sleeping with? Who will that person that you drink coffee with in the morning be?

Divorce reconfigures identity. It requires that women come up with new ways of seeing themselves and road-test them under grueling circumstances. For some people, this is a process of reclamation. Others, not having understood or accepted themselves to start with, get to a new place entirely. Women who don't depend on a man to define themselves won't find themselves in the predicament of Phyllis's friend, whose husband walked out after thirty-three years and who sobbed, "I have no identity now." They have come to look within themselves for a sense of self-worth, and much further than the fourth finger on their left hands. Although still bound by a network of responsibilities, women who end their marriages are no longer shackled to a relationship—and the value system in which it operated—that doesn't have their best interests at heart. Each of the women interviewed for this book has ended up with something very precious: a clear view of who she is, married or not. The process is painful but invigorating; in Nancy's words, "It just ripped those rose-colored glasses off my face, and that has been an *enormous* relief."

ROSEMARY

My regret is not that the marriage didn't work, but that it was based on people who didn't know who they were.

Without a husband's powerful voice in her ear, a woman learns to find her own path, to approve of herself instead of waiting for his nod or handout. For many women this results in a new sense of confidence in themselves and their ability to make good decisions. Amy, the single mother with lupus who lives in Oakland, was a confident person before she got married, but says, "It happens that the person I was married to very subtly and over a long period of time could really wear away anybody." Wincing, she recalls a point not long after her divorce, "and this is painful for me to admit, when I thought, 'What am I going to say? Do I have anything to contribute?' Martin wound up doing all the talking." Time has made the difference. "Now I want to talk, and I want to hear what I have to say. Not that I have the best things to say, but it doesn't matter."

ANNEKE
The most important thing is to be true to yourself. That's what I hadn't done for a long time, and it's a mistake I don't want to make again.

Women, unfortunately, often contribute to their own predicaments by resolutely closing their eyes to the mounting evidence that their husbands were neither willing nor able to give as much as their wives had been doing for years. They feel constrained by unwelcome choices, between being selfish or selfless, between responsibility to themselves and responsibility for others. Ultimately many come to reject that either/or view of the world because *it doesn't make sense*. The conflict between what they "should" want and what they do want—voice and equity—is resolved, and a tremendous burden unloaded. Self-esteem takes its place. As these women joyfully discovered, once they got to know themselves, they could begin to like what they found. This in turn enabled them to reject false choices, and to see their own interests and those of people around them more clearly.

Non-Wives Reject the Verdict of Selfishness

A set of "shoulds" with enormous cultural and historical authority reinforce the traditional wife's self-sacrificing behavior. Even if she can't quite measure up to that perky gal profiled in *Redbook*—that aerobics teacher who crochets and sells her own Christmas decorations while preparing a holiday feast for a homeless family and nutritious snacks for her own overachieving brood—she takes legitimate pleasure in her clean house, carpooled children, and cosseted mate. She's supposed to do for others, to accommodate (even if she can't resist carping incessantly about how selfish men are). Yielding to what Mary Catherine Bateson calls "the illusory clarity that comes from having laid self-interest aside" feels good. And when a wife accepts her position at the bottom of the family's hierarchy of needs, the message comes across: she *is* less worthy, less important, less deserving.

A major benefit of changing the job description from "wife" to just plain "woman" is the freedom to acknowledge—and indulge—desire. The selfless wife keeps the Pandora's box of unspoken wants and consequences shoved firmly in the back of the closet; peeking inside would be altogether too risky. But divorced women are no longer constrained by such taboos. It's quite a change, a little frightening, and definitely fun. Tory remembers going over to her stepsister's house after the woman's husband of fifteen years had moved out, and seeing a sign on the refrigerator door. "It said, 'Who's going to say no?' It was this freedom. It was so exciting to my stepsister, and I thought, 'Wow, I didn't realize that she felt that way all those years.'"

On and off during the course of her divorce Lorraine saw a counselor who was referred by a fellow teacher at her elementary school. At their final session the counselor urged her to "go out there and enjoy your life." Lorraine still remembers thinking, "What?!? Is that allowed? No, I'm just prepared to get out there and *live*. I don't think anyone gave me permission to enjoy." Until then, Lorraine had seen that attitude as self-indulgent, not on the menu of approved behaviors. For her, getting divorced was a means of jettisoning the arbitrary characterization of what is done for others as "good," and what is done for oneself as "bad." The habit of categorizing behavior in that way dies hard, especially for women who acted as their husbands' caregivers. In the thick of her divorce, when she had to change the locks and hold firm against her manic-depressive husband's demands, Eileen "had to keep reminding myself, 'You have to think of yourself first.'" It didn't come easily. Whether it means letting the dust bunnies accumulate or spending a Saturday in bed watching old movies, a woman is no longer governed by an abstract notion of "duty." She doesn't need to worry about whether her behavior looks selfish.

In the process of acknowledging her own desires a woman may

fear the loss of some vulnerable, generous, feminine part of herself. This is a legitimate concern; loving self-sacrifice fills a real need, and can bless those it touches. What goes by the wayside as women grow in this way, though, is not the ability to love but the impulse to give oneself away in the loving. What develops instead is the ability to be more discriminating about what is given away and to whom, so that it is a choice, not a habit.

Because a woman's self-sacrifice often encourages excessive dependence in others, newfound selfishness can work to good, often unexpected ends. For example, it can serve as a catalyst for those around her to take increased responsibility for themselves. Faced with the prospect of self-sufficiency, Karen's husband acknowledged the fact that he was never going to make ends meet as a freelance illustrator and became a licensed acupuncturist. My ex-husband went on to get the graduate degree he'd only dreamed about during our marriage.

The Painful Loss of Role

Women who end their marriages (and worse yet, who remain unattached by choice) defy category. They are what historian Carolyn Heilbrun dubs "ambiguous women," in contrast to those whose identities are anchored by their male partners. It hits home for me when I encounter a form with separate boxes for "single" and "divorced." I resent the distinction, and check both in the faint hope of irking some statistician down the line. The process of becoming an "ambiguous woman" took Tara nearly four years after Hal moved out, but, she says, "gradually I felt much stronger in who I was." Some women are ill-at-ease with this ambiguity and the way it may threaten or confuse people around them.

Confining though it is, the role of wife has its comforts, and Willa feels this loss of role acutely. Ironically, although she's a high-profile television producer, Willa's not burningly ambitious. With no false modesty she says, "I can't name any point at which this career was a conscious goal. I didn't have the drive. I just happened to find a career that suits my personality." Although the change in her domestic life has as much to do with her son's going off to college as with her divorce, Willa says sadly, "I have no role. That's the thing that hurts me the most: I have no domestic role. If I cook,

I cook for myself. And I still live like I'm married with children. I still have the house and the dog, I have all the stuff that a wife should have."

Ignoring the advice of family and friends, Willa is reluctant to sell her big house and move to a smaller, more centrally located place. "It's like I don't want to cut loose that part of my life," she explains. "It still defines me. I wanted to be a wife and a mom, that's all I ever wanted to be. I still don't see myself as a single lady with grown children, I really don't." But although she misses the companionship of marriage, Willa is not sorry she got divorced. "The miserableness of being by myself—or learning how to be by myself, 'cause it's something you have to learn how to do—that's less painful than being in a bad marriage, where home was not a refuge, the last place I thought about going. I'd rather run through a bunch of alligators."

Having gone from dependent wife to self-supporting artist, Winifred succinctly describes the change of role: "I've switched from being a service person to being achievement-oriented." The energy that went into taking care of Marty and making a loving home for his three daughters now goes into her own interests and career. One reason Winifred now feels good about herself is that, as for many other women, divorce made it acceptable for her to explore traditionally male roles and spheres of influence, in her case real estate investment and money management. The measure of her competence is no longer limited to the conventional caregiving domains.

Passed from father to husband, a woman is conditioned to accept and embrace the role of wife and helpmate. She is expected to take someone else's name and to put his priorities first—his mood, his job, his leisure time, his sexual proclivities—which is little rewarded in a culture that prizes individualism. The profiles in *People* magazine—and the big bucks—go to people who succeed in their own right. Unless you count winning the Mother of the Year Award, or having one's vacation home featured in *Architectural Digest*, rewards for achievement in the domestic realm are few. Living through another person may offer security, but the liberating truth is well put by Barbara Ehrenreich: "Roles, after all, are not fit aspirations for adults, but the repetitive performances of people who have forgotten that it is only other people who write the scripts."

Running the Emotional Gauntlet

Divorce brings enormous upheaval: logistical, social, parental, financial. Emotions follow suit, with moments of euphoria ("I *knew* I could get that job/change the oil/handle that snotty bank manager") followed by wrenching panic attacks ("How can I possibly handle the school play *and* my hateful boss *and* the holidays *and* that creep never calling back *and* the car payments?") and periods of reassessment. During a divorce both spouses behave badly at one point or another. Trust is betrayed, hopes dashed, and innocence lost.

How long and how hard these emotions hit depend on many factors, including what went on during the marriage, how long the divorce took and how much blood was shed, and the underlying mental health and resilience of the people involved. Time alone does the job, but solvency and support systems speed the process. Women with a supportive family and some money in the bank have an enormous advantage, but even those who went through it totally alone emerged saner and stronger than ever, though it took them longer to heal. During my divorce I was manic. I was wildly efficient at some points and barely functional at others. I knew I couldn't trust my impressions of people or events, but didn't know where to turn instead.

The roller-coaster syndrome is common, with some mood swings taking weeks and others hours. Writer Jane Shapiro described feeling "like two people at once—an old woman who has lost everyone and a girl whose life is beginning at last. I'm an ordinary divorcée, vivaciously mourning. I rise in the morning peppy and sanguine; only hours later, a heaviness grows in my chest and I'm near sobbing. I weep for hours, fall asleep still snuffling, then wake in the night with tears running into my ears. The next day, I'm shaky and refreshed; at midnight I can't sleep, because I'm excited, planning my moves."

Willa, whose sixteen-year marriage to a fellow television producer broke up two years ago, still wears her emotions on her sleeve though she's a well-grounded person. She advises women to be prepared to run a range of emotions and feelings. "Don't put too much pressure on yourself, because your emotions won't settle down for a while. And get a support system. Talk about it, *say* it," she advises. "I was so embarrassed it hurt, and then I would tell anybody that would listen that I hurt. I'd give a homeless man a quarter just so he'd listen." For Willa the period of intense turmoil lasted a year and a half, and two years later she says she still tells people, "I'm crazy, and one day I'll be sane. My emotions just run crazy, crazy." That's why Abigail Trafford, writing about divorce, titled her book *Crazy Time*, a period that she describes as starting when a couple separates and usually lasting about two years. It's a period of turmoil that has to run its course, an emotional process separate from the legal one that doesn't coincide with the serving and signing of divorce papers. The legal process can take six months or six years, but the duration of the roller coaster of "crazy time" is surprisingly predictable.

Fear: "Half the fear I'd had about leaving and being on my own was self-inflicted." Jodie

As discussed at some length in Chapter Two, the predominant emotion for most women when they end their marriages is fear—of financial hardship, of messing up their children, of loneliness and self-sufficiency. Jodie was afraid that her family wouldn't support her, Eileen that she'd never get free of her ex, Amanda that she was wrecking her child's life, Monica that she'd never get laid again. Not one of these fears was borne out, but they still had to be dealt with.

According to current theory, these fears are rooted in female psychology. Psychologist Carol Gilligan believes that men and women may experience attachment and separation in different ways, and that each sex perceives a danger that the other does not see—men fear connection, women fear separation. Autonomy, a male ideal, is completely contrary to the ways in which wives and mothers place themselves in the world. They define themselves in terms of their relationships to others, and when a relationship with an important person disintegrates, so does that element of a woman's self-definition. In forgoing the caregiving activities of marriage, in actually courting isolation, they jeopardize the very essence of who they are. How will they *matter* anymore, and to whom? No wonder it's so scary.

"This does not reflect well on who I would like to be, but my major emotion was fear," admits Celeste reluctantly. A year after her divorce the French teacher in Houston found herself without a boyfriend for the first time since her teen years. She had never learned how to live alone, and for two years was afraid every time she put her key in the lock, "not physically, just panicky about being alone." Celeste hid it well by being busy all the time, but one night finally had the courage to wait it out instead of calling somebody or going out. "I realized that I was afraid I was going to die, it was that kind of intense panic," she remembers, "but it only took that one time." Though those fears were acute, in retrospect they paled beside Celeste's realization that "no specific thing can ever be as bad as the generalized fear, terror, and anxiety that I lived in while I was married." Sometimes a moment with her new fiancé will summon up the years she spent being afraid to say the wrong thing. "I still feel that panic seize my chest," Celeste admits. "I had no idea I was doing that to myself all those years."

DELORES

A lot of decisions we make out of fear and doubt because of how we were raised. We think, "This is the model of the woman I need to be" instead of "This is the woman I am and this is what I am capable of."

Many women's fears center on loneliness, and they are often justified. Married for eight years, Amy headed into her divorce with a head full of fantasies, figuring that she'd get Martin out of her life and that everything else would stay the same. The loneliness was a shock, she frankly admits, although, she adds, "to tell you the truth, I was alone when I was with Martin, much more alone." Amy particularly misses her social life (he cooked; they entertained; she doesn't), and having a buffer between herself and her headstrong six-year-old. She describes a poignant moment when she spotted Martin and Cara playing at the park around the corner from their Oakland house and felt like "just slipping back in there as if nothing had happened, and the three of us carry our groceries home." Amy gazes wistfully out the window, then straightens in her chair and acknowledges, "I want something I never had. So it doesn't really matter, it's worth it." It does matter—Amy's loss is real—but confronting it is part of her rough road to self-reliance.

For about a year, fear predominated for Yolanda, the public relations director with the learning-disabled son. She worried not just about how she would provide for young Will and cope with Steve's rage, but about how she would handle day-to-day life on her own. Leaving her marriage was "an enormous act of courage," she says, "because I was very, very afraid of everything. But," she quickly adds, "my own inclination to be sane and to be a survivor came *roaring* to the front as soon as I walked out that door and had to believe that I could figure out how to buy car insurance. I had to and so I did." Post-divorce she feels far stronger, even though she now realizes that she needn't have been so afraid in the first place. "The fear subsided as I managed one unmanageable thing after another," she explains, "but as I did, I began to believe that nothing would come up that was really impossible to deal with, and that's when pure anger took over. I was very angry at my ex-husband, and felt kind of victimized." The anger gave way to "just comfort," but it took about four years for Yolanda to reach this feeling of closure and to acknowledge what she considers the greatest gift of divorce: "taking ownership of your life." Choosing to take responsibility for oneself is momentous, and it is only natural that the decision be accompanied by a welter of powerful, often conflicting emotions.

Anger: "I was so angry I felt like my veins were going to burst." Megan

Like Yolanda, many women I interviewed passed through a trajectory of emotions in which fear gave way to anger, often with surprising force and at unexpected moments. The succession is a natural one, familiar to any worried parent who finds herself furious when the kid turns up safe and sound. Anger is part of letting go of the past and adjusting to life after divorce. For women who were told they were stupid or incompetent during their marriages, a surge of hostility toward the men who made them feel that way is a common first step in regaining self-esteem. Anger also occurs among those who were submissive during their marriages and who now feel, and sometimes resent, the stress of having to become active in unfamiliar ways. In *Crazy Time*, Abigail Trafford notes the important psychological shift represented by the release of repressed anger. "The emotional ground rules immediately shift to self-interest," she writes. "Anger is a way to identify yourself as a single person."

For some women, like Willa, anger comes foremost. At the time of her divorce, the producer describes herself as "mad as hell." Then she stops to think, leaning across her wide walnut desk. "You know, I think I've come almost full circle. Two years ago, I would have said, 'That son of a bitch, he didn't support me, he didn't help me, I think he's jealous of me,' those kind of things. But I've come to understand some of the dynamics of the relationship just a little bit better. And there's not so much pain. I can even see my participation in it. Before I couldn't." Two years ago, she says, "I blamed everybody. First I blamed him. I blamed myself. I even blamed my son, in a way." Now she feels blame is beside the point, and many women join her in that sentiment.

For Natasha, separation resulted in a period of total turmoil during which anger ruled. She was furious that Florian had come back to Santa Monica from a lengthy East Coast directing job with a much younger woman in tow, furious that he had left it to mutual friends to clue her in, furious that her whole world had been turned upside down, and furious that he didn't understand why she was so furious. Even in the depths of her rage, though, Natasha understood that her anger needed constructive outlets. A dancer and singer, she worked consciously to release it in different ways, taking dance classes, exploring her feelings, and talking to people. In addition to a day job, Natasha took freelance work and taught at night. Also helpful as a counterbalance was her euphoria at coming out from under the stress of the marriage, "being free, for the first time, of his driving presence," as she puts it. Three years have now passed, and she says serenely, "I think I'm just coming out of it." In her marriage, Florian did the talking; now her natural garrulousness is back.

At their angriest many women cast about for people to blame. Ex-husbands come in handy for that purpose, though most would admit that "it takes two to tangle," as Francine put it, though she still points a finger at Frederik's less than stalwart behavior. Though they can hardly be considered objective, most women felt that they had made much more of an effort to communicate during the marriage and to work out problems as they materialized. "I don't hold [the failure of the marriage] against my ex-husband, but I blame him," says Eileen, "because I think that there were many, many opportunities to change things." Hillary casts a wider net. "I blame society. The way girls were reared, the way I bought into all those societal notions of what the good girl was."

Understanding the source of the oppression does not resolve the psychological deadlock. There's a big difference between grasping how one person comes to take advantage of another and dealing with the emotional effects of having been exploited or abused. Women who give vent to their anger are far outnumbered by those who repress or deny it, as evidenced by how many of them placed great value on remaining amicable with their husbands during and after the divorce. Some did so for the sake of the children, but justifications like Phyllis's—"I don't like confrontations"—were extremely common.

After all, anger isn't ladylike. It turns men off and unnerves other women. "Above all other prohibitions, what has been forbidden to women is anger, together with the open admission of the desire for power and control over one's life," concurs Carolyn Heilbrun. No matter how grievous the insult, women are expected to respond with patience and forbearance. But anger serves a purpose; it can be a catalyst for growth. The challenge is to use the energy it releases constructively, and to move on. Otherwise women are consigned to bitterness or vindictiveness, which are wasteful and tiring. If the anger is turned inward, it is manifested as depression.

Depression: "I think I slept for about six months." Winifred

When women turn their backs on the glossy collective fantasy that packs the pages of *Bride's* magazine, there's a price to pay. Divorce no longer carries the breath of disrepute it did a generation ago, but it is still censured. "Intended or not, divorce punishes divorcing men and women," writes historian Glenda Riley, going on to point out that "because women, the long-time protectors of home and family, fail to hold crumbling marriages together, their misfortunes seem particularly appropriate." Women are quick to accept those misfortunes as deserved: toss off the mantle of self-sacrifice and you'll find it's cold out there.

In an essay on her divorce, writer Ann Patchett harrowingly describes the despair that accompanied the end of her marriage. She reflects on the smug, unthinking cruelty of people who have been lucky in life and who equate their good fortune with personal goodness. Working in a fast-food restaurant, wearing a funny hat, and serving fajitas to one-time classmates was horribly painful, but, she admits, "I did not die." Hitting the bottom of the emotional barrel freed her. "I came to see that there was something liberating about failure and humiliation. Life as I had known it had been destroyed so completely, so publicly, that in a way I was free, like I imagine anyone who walks away from a crash is free . . . I knew there was nothing I couldn't give up." Though it's a lousy way to learn, depression serves as a prelude to grief and a necessary part of the mourning process.

I didn't sink as low as Patchett. But though I seldom miss a meal and refuse to hide behind my answering machine, there were periods following my separation when I was too demoralized to eat or to pick up the phone. "Is Ashton all right?" my mother asked after the first holiday kid swap with my not-yet-ex-husband had reduced me to an all-day sobfest at my sister's kitchen table. "No, Ma," replied my sister, "she's a fucking mess." My psyche turned a corner in the back seat of a cab during that raw, exposed period when I would have discussed my intimate circumstances with a fire hydrant. "Doesn't it feel like a failure?" asked the cab driver (far from the first with that cloddish question). "No," I replied honestly, drawing a deep breath. "It's not a failure, it's a *loss*."

Nancy's depression expressed itself as disappointment, which colored the whole period for her. "Somehow he thought that he was so above it all, blaming it all on me instead of trying to rise to the occasion," she says. "That was my biggest real disappointment, his saying, 'It's all your fault.' That *hurts*. It's incredibly painful." Contributing to Nancy's sadness was disappointment in herself, that she had let her marriage erode her self-esteem so deeply and now had so far to go to extricate herself from this web of illusion and self-delusion.

Megan, the woman who works for a cable-TV network in Boston, manages four words to describe going through her divorce: "Devastation. Mourning. Loss. Death. The pain is as real as losing anyone you truly, truly love in death. You mourn it the same way, you're scared of it, you're paralyzed by it. And you are for a long time." But she absolutely does not feel that her marriage was a failure, or the divorce a disaster, but rather a lesson she had to learn. "I feel if anything it was a trampoline to growth." When she first moved out, grieving deeply after years of struggling with the decision, Megan felt very shaky. Things began to change "when I realized that I was going to survive." She remembers a friend saying at the time, "'I promise you three months from now you're not going to feel like this.' Three months from that day we went out to dinner and I laughed with her. I said, 'I feel so much better, so much more stable, more sturdy on my feet.'"

Divorce is a time during which women's traditional sources of identity and self-esteem fall away and new sources have yet to be established. It's no surprise that what Tara christened "post-departum" depression is common. The treadmill many single mothers find themselves on makes them particularly vulnerable to it. Struggling to feed and clothe their children, they have little time or energy left over for the luxury of looking after their own needs. Those whose former husbands have little contact with the children are especially resentful. Reluctant to show anger in front of the kids, and perhaps unwilling to confront its source, many mothers turn it on themselves.

But depression is a transitory state on the path to resolution. Most women in this study did hit the bottom of the barrel, and quite a few took years to climb out. But whether despair peaked before, during, or after divorce, sooner or later it yielded to closure and contentment. In her study on the long-term effects of divorce, the nation's largest such study, Judith Wallerstein found a significant amount of depression, especially among wives, *before* the separation. She also found that those who got severely depressed afterward were not the ones who had been depressed beforehand, and were seldom the initiators of the divorce.

How Long It Takes: Mourning and Closure

In her classic book about death and dying, Elisabeth Kübler-Ross identified five stages of grief: denial, anger, bargaining, depression, and resolution. Divorce is a death of sorts, and people going through it pass through these stages as well, though not necessarily in order or for a predictable amount of time. It took Violet twenty years to put her marriage was behind her. On the other extreme is Rosemary's confession: "You know, this is an awful thing to say, but the day he moved out I felt like I'd never been married."

Mourning During the Marriage: "I knew for so long that it was over." Anneke

Many women who end up initiating their divorces have already spent months or years deliberating the consequences, passing through the stages Kübler-Ross describes long before lawyers come to mind. As Anneke, the romance writer from Tucson, puts it, "I knew for so long that it was over." A friend of Anneke's is a Native American medicine woman, experienced in a practice called soul retrieval, which is based on the theory that part of the soul splits off when a bad thing happens to someone. "The shaman goes on a spiritual journey to retrieve these pieces," explains Anneke, who had several of these soul retrievals done. "It was an extremely powerful experience for me, and after that things really started changing in my life." Once the divorce was in progress, Anneke's primary emotion was relief, and that feeling hasn't changed. "Sometimes I have a bad day," she admits with a smile, "and then I think, 'Oh, count your blessings, you could be back there.'" Closure took longer. A shared passion for opera and classical music had really connected Anneke and George, and buying and listening to new CDs (he kept the record collection) filled her with grief. Now, two years later, "that's gone," she declares. "I can sit down and listen to things that we used to listen to a lot together, and I might think of him in passing, but not with any deeper emotions."

Virginia was one of those women whose period of greatest stress preceded the decision to divorce, when she worked full-time, paid the bills, cared for the baby, and began to wonder just what her husband was there for. She mourned her marriage while she was in it, "not once I was out of there. Are you kidding? I had dance parties, I got myself a new stereo, I had a *blast*. I had a party on the day we signed, a party when it was official in the court, a party when I got permission to leave. I got new music every time, we danced up a storm. I just celebrated." The initial euphoria has given way to a more realistic appraisal of her circumstances, but despite the disadvantages of being a single mother, Virginia says unequivocally, "I would certainly choose it over marriage to him or marriage to the wrong person."

Women who mourn during their marriages constitute part of the group described in both Trafford's and Judith Wallerstein's research whose spirits improve following the decision to end their marriages. Trafford describes them as "usually women who had been seriously depressed in a difficult marriage and now had reached the painful decision to divorce, often with the help of psychotherapy." Some, she goes on to say, "in fact seem to glow after the breakup and continue through Crazy Time with feelings of having a second chance at life."

Achieving Closure: "The hardest thing is the death of the dream." Olivia

In the throes of the worst pain and anger, a woman may wonder just when, if ever, life is going to start looking up." Trafford quotes counselor Sharon Baker of the Los Angeles Divorce Warm Line: "It depends on how intense the relationship was and how big a need it filled; how self-sufficient and mature a person is; and if you have the three good fairies of luck, money, and health." The more stable and psychologically healthy a woman is going into her divorce, the sooner she is likely to attain equilibrium afterward, though financial stress or a hostile ex can certainly slow things down. Divorce is consistently ranked second on national lists of stressful life events, after the death of a spouse, and on one it ranks first. The consensus among researchers it that it takes at least two years to regain emotional equilibrium after a divorce, and that was the period cited most often by women in this study.

Olivia concurs with the two-year average. Despite always having thought of herself as perfectly capable of being on her own, she was surprised how lonely she felt after getting divorced. Late nights spent in her shop got her through the lonely months immediately afterward, and Duncan's speedy remarriage was painful but therapeutic. And when her business partner was widowed a year later, Olivia ended up moving into the woman's comfortable house. The arrangement suits both of them, now in their sixties, extremely well, and Olivia is very content. "I like myself much better single," she says with a laugh, "because I'm standing on tippytoes, right? I can *be* who I really am, I'm not always trying to be somebody else."

Duncan was a terrible stick-in-the-mud, and part of Olivia's well-being can be attributed to a sense of physical liberation. "I felt like a cork that had been held down in water and was suddenly let go—I popped right up," she says gleefully. Yet she still has moments of grief, especially around the holidays. "I think the hardest thing—I'm going to get teary now—is the death of the dream, what had become of this marriage," Olivia maintains. Moments of nostalgia are inescapable, but the cure is swift. "Then every time I'm with him, I think, 'Oh, God, I'd forgotten that [negative thing about him]. Thank heavens I'm not there anymore.'" Dina, the immigration lawyer from Sacramento, also has moments of doubt, though they are less and less frequent. "It's awfully hard sometimes with the kids to be around a lot of intact families," she notes. "I think, 'Gee, maybe I should have worked harder, maybe it would have worked, maybe he would have changed.'" But the cure is the same: "all I have to do is see him or talk to him, and I'm over it, it's really true."

Researchers find that three-quarters of divorcing spouses experience some lingering attachment to their mates after separation, and that those who don't are usually those who went through the psychological separation process during the marriage. There are always moments when the grass looks greener. Ambivalence is natural, especially in the immediate aftermath of the decision; over time, it turns to nostalgia, a feeling inherently rooted in the past. It testifies to how much their marriages and their mates once meant to these women, and sometimes still do.

When Closure Takes Longer

Feelings of attachment may linger, and while it's important not to get bogged down by them, denying them can impede emotional progress. Karen, the sculptor, thinks maybe she'll feel a sense of closure when she has another relationship. Sharon didn't feel closure until six years had passed, when her ex-husband Charles initiated a Christmas visit. Another factor was that Sharon no longer felt the need to offer any excuses or justification for what she had done. "We were finally able to communicate as intelligent friends again, as opposed to ex-husband and ex-wife." Happy that the original kernel of friendship has survived, Sharon in retrospect thinks that marriage was a completely inappropriate venue for it, and that she and Charles "should have stayed best of friends."

Though long established in her career as a cookbook writer, Wendy didn't feel a sense of closure until several years ago, thirteen years after the divorce. That's when Frank, her cardiologist ex-husband, got married for the third time, to a woman who Wendy says happily is absolutely perfect for him. "Now that he's in the hands of a competent woman I don't have to worry about him anymore. Brenda [the new wife] has been the best thing that ever happened to me." She laughs.

Just as self-reliance is crucial in building self-esteem, self-esteem is necessary to achieve closure. Until a woman feels good about herself, she may well stay stuck in anger, bitterness, and depression, or at the very least take much longer to come to terms with her divorce. Self-esteem was completely missing from Violet's marriage, and physical and verbal abuse took such a toll that it was a full twenty years before she was able to put the marriage behind her. "I used to be like a really soft person," she says. "Through my experiences going through the marriage and with the divorce, I got stronger, and I think it's for the best. You can't be too soft, because people, men and women, will take your kindness for weakness."

Work was the key, giving Violet the courage and the means to rebuff her ex-husband's custody threats and to begin to see herself as a competent, worthwhile person. She's always worked with children, and now provides day care in a shelter for battered women. When the shelter was featured on a television show about domestic violence, the director put Violet on the air.

I was nervous to be on the show, but after I talked and explained to everybody that was listening that there was a place for women that were having problems with domestic abuse—and do not be afraid to call that number and make that step and go, get out of your situation—I felt good. I felt like I was helping some woman out who was sitting there crying and saying, "I don't know what I'm gonna do," same thing I was saying. If she really listened she'd say, "That woman's telling the truth," because I really was.

It was at that moment that Violet felt the door of her horrendous marriage finally close behind her.

What Helps and What Doesn't

The cardinal rule for coping with the stress of transition is talk, talk, talk. Used to sharing their feelings, in this regard women have a tremendous advantage over men, though that doesn't make it easy for all women. "I was ashamed. I was embarrassed," says Violet, like many abuse victims, "and I figured the less people knew about my private life, the better. I kept it all in, and it was really hard too." She chuckles as she says this, not a trace of self-pity in her voice. "If I had to do it over again, I'd talk to somebody, because it takes you longer to get through something if you have no one to talk to." Movingly, Violet has gone from being someone unable to take advice to someone in a position to dispense it.

GETTING HELP FROM THERAPY AND SUPPORT GROUPS

Therapy and support groups can provide invaluable guidance before, during, and after divorce. (See Chapter Two for information on how to find a therapist.) Advice and companionship come in all shapes and sizes: the guy at the next desk, childhood friends, Al-Anon, Mom and Dad. Less important than who is doing the listening is the catharsis of articulating feelings and rationales, but it's certainly a plus when the listener is a trained professional. Providing an objective viewpoint and moral support, therapy helps many women to see issues more clearly, and guides them through the turmoil of the divorce and its aftermath. Forced to take a longer view, they become able to take it on faith that their pain will diminish and the rewards grow clearer.

Francine remembers asking herself during the worst of it whether she would have done it had she known how bad it would be. "*Now* I look back and I say 'Of course,' because I went through it. But when you're in the middle of that dark tunnel, you wonder, 'Why did I do this? This is terrible.'" Her therapist was a big help, especially in assuaging Francine's fears about meeting new men. "My therapist even said, 'I think you're going to have the opposite problem,'" Francine recalls with a smile—and she was right.

For Sharon, therapy offered not just perspective but the reassurance that "I didn't have to settle for this, that this wasn't the way it had to be, that I had some level of control. She told me that I could really choose my destiny." As her marriage progressed, Sharon began traveling a lot and working late. Her marriage paid the price, and she doesn't want to repeat the pattern if she marries again. She remains extremely career-driven, but with the help of her therapist she's made a real effort to integrate her professional and personal lives. Sharon sees now that "the most important thing is a sense of personal confidence and self-respect, which were very much lacking when I was married."

Like Francine, I began therapy about a year and a half before my marriage ended, and continued for another year thereafter. As for Sharon, therapy was instrumental in helping me to avoid some of the mistakes I had made in my marriage in my new relationship. Yes it was swell that I had met this wonderful man, my therapist would agree in the driest of tones, but no, it didn't follow that my life would be an utter wasteland if he hadn't happened along. It was demoralizing, but extremely educational, to realize how eager I still was to sell myself short.

A good therapist guides the patient through the gauntlet of emotions triggered by the severing of an important bond. Pain is part of the picture, but as marriage counselor and clinical psychologist August Y. Napier points out, "almost all clinical growth takes place in the imperative of unhappiness." Therapy helps people at every point in the process to sort through their feelings and to understand even the most extreme ones as normal and potentially constructive.

ANNEKE
I would say get as much help as you can from wherever you can, whether it's from friends, from therapists, from books.

GETTING HELP FROM FRIENDS

Almost every woman spoke of a best friend, or a network of friends, usually women, who helped them through the rough times. During the seven interminable months my husband and I still shared a roof after having agreed to divorce, we arranged to spend alternate nights and weekends out of the apartment. Compelled to leave, I would take refuge on various friends' sofas and sob.

When Tory, the human resources coordinator, got divorced in the early 1950s there weren't many other single mothers around. She says she felt like the woman in the old song "Harper Valley PTA," scorned because she was a single mother. Estranged from her family, Tory couldn't have gotten through those early, lonely years without the unwavering companionship of another woman in the same boat. They assembled toys on Christmas Eve together, they attended each other's children's graduations, they were the parents. At one point her friend said to a guy she was dating, "Why can't you be more like Tory?" Tory recalls with a laugh. "Patty and I would say, 'If we could just get through this sex thing, we'd be very happy!'" She says, "She was just a great partner. That was my support."

In the wake of divorce friends sometimes hasten to ally themselves with one half of the couple, or to distance themselves entirely. Memories of friends who failed to come through were sharp. Certain friends got Nancy through the day-to-day stuff, but "some just steered way clear, and that was incredibly disappointing to me. They were married, and if anybody's having any kind of conflict at home, they just don't want to hear about it." Observing the same syndrome, Rachel, the sales clerk, offers a wry diagnosis: "They think that's it's a disease that'll spread and they don't want to get sick."

Often friends are more judgmental than family. A strong and resolute person with a thriving practice as a massage therapist, Tara nevertheless felt punished by some of the people she knew. "A lot of the friends that Hal and I had really became Hal's friends," she recalls sadly. "I have a very strong support system among women, but I lost all the men. And that was very painful." In retrospect, she says, "What I didn't understand is that I would really be alone." Relations with her closest friend, who could barely speak to her own ex-husband twenty-four years after their bitter divorce, are still strained. "She wanted me to be really angry with Hal, she wanted to feed any rancor," explains Tara sadly, "and I didn't feel that way."

Some women chose to end friendships of their own accord. Celeste found that she didn't want to be friends with her husband's male friends anymore. "I think they really had liked me, and they were a little bit surprised and disappointed when I said, 'Brad really needs you now, more than I do, and it's going to be difficult for him if you stay in touch with me.' They took that as a brush-off, which I suppose it was, in a way." Nevertheless, making the break was right for her.

Sometimes there's a scramble among friends to establish the "good guy" and take sides accordingly, which usually places the initiator at a disadvantage. Phoebe, the clinic administrator, lost all of her and her husband's once-mutual friends. "Howard contacted all of them and shared his story, said he didn't understand why I was breaking up our family and how he wanted me back." A very private person, Phoebe did not attempt to tell her side of the story, feeling it wasn't anyone else's business. Wendy was similarly ostracized by her conservative community. She and her ex-husband Frank's mutual friends, all doctors and doctor's wives, stuck with Frank; he was the doctor, after all. "I suppose that's predictable," says Wendy with a nervous laugh, "and some of them have since made overtures. Some of the women especially have said, 'My God, how did you have the courage to do it?' and admitted a little envy. But they're all still married and now they're too old and they're not going to do anything about it." Wendy's life is full now, and she doesn't miss any of those people.

Ginny, the art director in Rochester with the alcoholic ex, was another who found herself cut off from friends with whom she and her husband had always socialized as couples. She says cavalierly, "I just looked at them, and thought, 'Oh you poor people, it's too bad to learn that about you,'" It hurts, she admits, but she also has plenty of friends to whom her single status doesn't matter. Key among them are the two friends, her "Friday night people," whom Ginny has known and skated with all her life. An unexpected bonus of the divorce was ending up with a really good relationship with her ex-mother-in-law, who got a taste of what Ginny had been dealing with over the years when Mike moved back home. Ginny has no illusions about where her mother-in-law's ultimate loyalties lie, but appreciates her help with the three children.

GETTING HELP FROM FAMILY

Many women anticipated that family members would react to the news of their decision to divorce with guilt trips or finger-wagging, but were pleasantly surprised. Theresa's family supported her completely—after a rough start. Theresa's mom was going through her own divorce at the same time, and was convinced that that was why Theresa had decided to leave David. Finally Theresa, the nurse with three young daughters, took her mother by the shoulders and declared firmly, "Ma, I don't hate my husband because you hate your husband. I hate my husband because *I* hate *my* husband." That set her mother straight, and she supported Theresa from that point on.

Not only did Phoebe's parents take her and her baby in and physically shield her from her abusive husband, but they were an unwavering source of moral support. She needed it, since Howard obstructed the divorce in every imaginable way. A petite African-American woman whose close-cropped hairstyle emphasizes her delicate features and graceful neck, Phoebe shakes her head as she recalls that grim period. "We were in and out of court so often," she says wearily. Her father was her mainstay. "For those two years, any time I went to the attorney's office my father would take off work. He would never say, 'I'm going with you,' he'd just be standing at the door, dressed, when I was ready to go." Her family also watched her son during the years when Phoebe completed her schooling and worked her way back up the economic ladder into her present administrative position.

Sadly, though, many families are just not capable of dealing with these kinds of problems, even at a family member's time of greatest need—when he or she is broke, divorcing, coming out of the closet. Sometimes they lack the emotional or logistical resources; sometimes it touches a loss of their own; sometimes they're afraid. Monica describes her family as "predictably indifferent to my plight. I could sort of see the circling of the wagons—'God, what if she asks us for money'—and there was no support."

Susannah's parents were great at moments of crisis, but their counsel at other times tended to take the form of reminding her just how much better off she was than most women—which is supposed to make you feel good but doesn't, as Susannah points out. Her sister-in-law was wonderful, though. "She was the only person, the only person of everybody, who said, 'This must be really hard for you.' Because everybody just assumed that if I left, it wasn't hard." Nancy's parents were deeply concerned, but their support tended to take the form of hand-wringing. "It wasn't 'You can do it, get out there!' it was kind of like, 'Oh dear, what are you going to do now? Oh dear, oh my.' And that isn't what I wanted to hear."

Most helpful to Yolanda, the public relations director, was not a person but a "pearl of wisdom" from a friend. "She said very simply—and it was an utter revelation, it's now something I try to live by—that it's not whether you win or lose, it's that you don't play the game. It changes everything. Disengage. It turned out to be critically important to me in getting out of all the stuff that was still cross-firing, crackling in the air after the separation between the two of us." In part because she had only one surviving relative, Yolanda made the decision to create what she calls a "family of choice," independent of legal and biological ties. "My friends are my real family," she explains. "My son has a grandmother of choice. So do I, and a brother of choice. They were all quite supportive and so were others, except for a couple of friends who were quite attached to my husband as well. I let him have them."

Wonderfully adaptive, Yolanda's behavior also reflects one of the ways in which the American family is evolving into a much more broadly defined social unit, despite the religious right's insistence that biology alone defines a family. This broader net is a great source of support and companionship for those whose relatives can't or won't come through.

For some women, friends, family, and church all overlapped as support systems, with varying degrees of effectiveness. In Beverley's case, because her father and husband were both ministers in the same church, family and faith were especially intertwined. "I had three really good friends, and I thank God for them," says Beverley, "because there was a point where my parents were too angry to help me and my friends were there in a big way." Beverley moved back home for a while with her baby, which proved very stressful. "I was in a lot of pain and I think that they didn't know what to do with that, how to help me," she observes. "It wasn't like falling off my bike, and that frustrated them no end." Behind the frustration lay the fact her family was very angry with her for leaving not just a marriage, but marriage to a minister. "It ended up being very messy because we were in a small community where everyone knew us," explains Beverley. "Especially the black Baptist churches, when a minister and a minister's wife are no longer together it is *her* fault, no matter what he has done."

GETTING HELP FROM RELIGION

Luckily, Beverley was able to separate the unsupportive behavior of her family and fellow churchgoers from her faith in God, which sustained her as it does many women when they forsake the security of marriage. She did stop attending services, because she didn't want to be around those particular people and she had nowhere else to go, but she continued with her own prayers and devotions. "I think if I had given up my faith, it would have destroyed me," she says bluntly.

When she moved back to Roanoke, Virginia, from a two-year stay in Shreveport, Louisiana, after her divorce, Beverley found a new church, which she describes as "a great place for healing," she says. Three years had passed, and now she knew how to avail herself of their help. "One of the first things I did was let them know what I needed, and they were wonderful. I was at a point where I had to speak up, because nobody was just going to notice. I felt myself kind of slipping. I felt like my life had been taken." Beverley is profoundly grateful that she didn't turn her back on her faith or her background despite the shabby treatment she received, "because it would have taken away everything in my life. That turned out to be what got me to where I wanted to be, even though there are times when I feel I'm not there yet." A better job and the possibility of home-schooling her son seem like distant goals, but, she says with a good-natured shrug, "you never know, maybe I'll think of a way."

Phoebe, like Beverley, found that some members of her religious community were less than supportive in the short term. Two weeks after moving out, a deeply demoralized Phoebe went to church with her baby in her arms. After the service the deacon came up to her and said, "You're going to hell."

I was like, "Excuse me?" He said, "I hear that you've left your husband." I said, "Yes, I did." He said, "Well, you're going to hell." I just busted into tears, because of course you like to feel that God is on your side, and then I said to him, "Well, if I'm going to hell, I'll see you there." I turned and walked away and tried to compose myself.

Fortunately Phoebe was able to see this cruel and arbitrary judgment for what it was. Always a faithful churchgoer (something her husband had mocked her for), Phoebe continued to attend services and to draw strength from them even in the bleakest times.

Many women prayed for strength during those grim periods, and for guidance that they were making the right decisions. Although Marella, the mother of five from Trinidad, is extraordinarily resolute, she felt great sadness when she moved out, symbolized by missing her own kitchen. "I was happy in my aunt's kitchen, but it wasn't my kitchen. I felt like I had lost something. I felt terrified at times, of being alone. Everyone saying, 'Are you crazy? What are you doing?' I felt, 'Gee, what if they are right?'" She's not sure those feelings were justified, "but they were real." The grief lessened after six months or so, and in another six months Marella landed a good job and turned the corner emotionally. "I said, 'Yippee! I'm going to do this.'" Financial crises were frequent, but Marella believes in the power of devotion and positive thinking. "It was very difficult," she says, but "through prayer I would be able to stay alive, to stay afloat."

Crucial in getting through the worst of it for Hannah, the mother of four from rural Wisconsin, was a 12-Step recovery group that she describes as an "overcomers outreach for adult children of dysfunctional homes." The program's reliance on a higher power integrated well with her own fundamentalist Christian beliefs, and Hannah called on both. The sense of relief she had felt when she made up her mind was swept away by grief when she actually filed the papers. "That was very difficult. I had very strong feelings of wanting to take it all back, even," Hannah recalls, her blue eyes narrowing. "It was like, 'Oh my God, what have I done? The kids are all in pain, Ed's in pain, we're all in pain.'" One day in the throes of her divorce she called a minister friend who belonged to the group and who gave her a "word picture" that carried her through the roughest time.

I told him that I just wanted to put everybody out of their pain, close my eyes to all of the problems. He said, "Hannah, I see you as being like the children of Israel. God led you out of Egypt, led you out of slavery, and now you're out in the wilderness, and it's really scary out there. It doesn't feel good." He said, "If you look back, like the children of Israel, you think, 'We're in so much pain. At least we had these certain comforts.' But you just have to keep going and know that the promised land is coming." I said, "That's right, that is *it*. And I just kept that picture in my mind.

A few months later Hannah got a call from a Christian friend who was very concerned about her. "She asked me, 'How is it, are you okay, how is your life now that you're divorced?' I said, 'You know, until now I just never knew life didn't have to be so hard.'" Now she sees herself as a happy and optimistic person, whereas "in my marriage for quite a few years I felt like I was just surviving. And life is about more than surviving."

Tara too found that spirituality was crucial to getting over her divorce, though her religious beliefs are less mainstream. In the preceding ten years she had done a tremendous amount of personal work, vision questing with Native American teachers, and "muscle testing," a way of accessing the body's wisdom. Her community of teachers in Boulder provided tremendous spiritual, psychological, and emotional support. What stands out the most for her during this period in her life, Tara says, "is my reconnection to myself, to my larger parents, to my nurturing by the universe, by the earth. To my freedom from the shackles of mundane relationships. Which was terrifying, because I let go of so many all at the same time. I lost friendships, and my marriage, and my lover." That didn't mean that being single held no pleasures. "Oh, I loved it. I *loved* it," she says delightedly.

THE COST OF NOT REACHING OUT

AMY

If you don't talk to people, you're isolated and scared, and it's going to hold you back.

Women who kept quiet suffered more and took far longer to recover emotionally from their uncoupling. Willa and her husband had worked in the same room with the same people, their best friends, for sixteen years, twenty-plus in her husband's case. "The most dignified thing we did," she says proudly, "was that no one in that room knew that we were having trouble, or even getting separated, until *after* I left. And that was tough. We had no support." She put her emotions on hold for a full five months before calling on her friends for the sustenance she so badly needed. "I was angry, and sad, and lonely," she admits, as well as tired of keeping up the Ms. Superwoman front. To people who dismissively told her, "You'll do just fine," she felt like snapping, "Dammit, don't just assume that because I don't complain that this isn't killing me!"

Some women may reveal their turbulent feelings to friends and family, but try to shield their children from any awareness. Amy took it a step further, working almost obsessively to keep her four-year-old's world intact during the three-year legal battle. Not a soul knew she and Martin were getting divorced, and in retrospect Amy vehemently wishes she had sought out comfort and support much sooner. "There was nothing to protect, nothing to hide. There's no disgrace." Feeling that she needed to be strong, Amy didn't shed a tear the whole time, "but now I'm starting to mourn, and it's hard." At the time she thought her stoicism was courageous, but she now says, "actually I think I was a coward, because I kept trying to protect this little unit so that nothing would change too soon. I actually see it in the reverse. I feel strong now, I feel braver now. I feel great that I did what I did."

Some women's reticence is rooted in shame. Eileen comes from an alcoholic family, and when her husband started drinking she was deeply embarrassed at finding herself in the same situation. "It's very hard to ask your family for help, even though they would, because it's kind of like, 'How did I do this? You've gotta be kidding!'" Eileen didn't want to tell people at work either, "to have people say, 'Oh, poor you,'" a reluctance she thinks is also rooted in having grown up with alcoholic parents. "You don't want sympathy." It made it harder. Eileen says, "I'm much more comfortable with myself [now]." Work is important to Eileen, but a lesson learned at Al-Anon, which helped a great deal when she was laid off from an earlier job as a sales rep, is that "I don't define myself by my work. I define myself by myself."

Another aspect of growing up in an alcoholic family is a reluctance to look to the future, but Eileen feels that changing. "I'm

WHAT ARE YOU MOST GLAD THAT YOU DID THE WAY YOU DID?

SHARON

I'm absolutely glad that I married my first husband. He was the right guy to marry, the right companion for learning the things I've learned.

GINNY

I'm really glad that I gave everything that I possibly could have to try to save the marriage, although there were times that my actions were so unexplainable and irrational and sick.

HANNAH

I am glad that I had the courage to finally say, "Enough." There were many times when I would think about it, and even begin to act on it, and it was much too scary. So I am very thankful for having that courage, because it's no small thing to have.

ANNEKE

I'm most glad that for myself I was able to end the marriage without real bitterness. I was just able to forgive him and say, "Okay, it's over. We had good times and we had really bad times and I learned a lot of really important things."

SOPHIA

I am most glad that I didn't resist him when he raped me. I have my daughter.

YOLANDA

I'm so glad about the whole thing. I don't think I ever could have begun to grow up, not in that marriage. Now I am not only grown up but growing up. It's obviously an unending process, and I'm enjoying it enormously. I think these are wonderful years for me, and I'm very happy.

much more optimistic now," she says. At the end, when she realized she wasn't responsible for everybody in the world, she felt tremendous relief and lightness. "Then you can do anything," she says, "and what's great about it is that you can't go back. Which to me is the biggest gift of all: you cannot go back."

How Women Feel After Divorce

Once she has run the emotional gauntlet, once the "emotional divorce" is behind her, a woman becomes open to a whole array of positive feelings. In the Children of Divorce project, Judith Wallerstein confirms that divorce resulted in positive life changes for the majority of women; interestingly, this was not the case for men, perhaps because fewer men were the initiators. Women's good feelings are rooted in authentic self-knowledge, the transformation that lies at the heart of the divorce process. Despite the hurdles posed by psychology, maternity, economics, and the expectations of others, women who end their marriages come into their own, for keeps.

They're Proud of Themselves

WINIFRED
I'm in charge of my life now, and I realize, "I've done this. I've created this, this life."

Pride itself is a victory, requiring that a woman rebut the stigma of divorce and overcome the sneaking suspicion that pride might be a selfish or inappropriate emotion. Susan Brownmiller notes that, as with anger, "exulting in personal victory is a harshly unfeminine response." She goes on to say that "the merest hint of gloating triumph—'Me, me, me, I did it!'—is completely at odds with the modesty and deference expected of women and girls." Phoebe sees no call for modesty: "It's been hard, but I've made my dreams come true, nobody but me," she declares firmly. Despite some really lean years after she was laid off and could find only minimum-wage clerical work, Phoebe is dead clear that the hardship was worth it. She got her college degree, she got raises, she got her son through high school, and she even got remarried last year. Phoebe is now the administrator of a thriving women's clinic in Bowling Green, Kentucky. "Believe in yourself and you can do it," she urges, holding up her own life as an example. "Now I have my dream job: I've excelled."

Helen delightedly describes herself as a "surly woman. Her plain face is split by a grin as she adds, "I have to be surly at home too, yes, I do. I have to stand up for myself. My husband's a domineering person and an engineer, so he knows the best way to do everything. Last night on TV someone said, 'You're the bravest person I know,' and this person says, 'That's because I want you to think that way.' And that's me." This feisty woman is a far cry from the one who cowered behind her living room curtains, trying to ascertain her husband's mood from the way he walked up the driveway.

"I'm very proud of myself, I have to say," Francine admits shyly. "I'm proud of how I handled it with Frederik, that I tried to be as compassionate as I could, and that I knew how to get help myself." She too has come a long way from the newly single woman whose eyes would gravitate to the answering machine "and those zeroes would have a profound effect on me, in a way they don't now." Going through a divorce has made her more cautious, but along with her awareness that things may not work out as planned, Francine's "confidence in myself as a person, in my judgment" has blossomed.

They Relish Their Independence

GLORIA

My daughter and I were clearing out the garden and she goes, "Oh, Mommy, can't that man from next door pull these bags out of here?" And I got so mad at her. [Laughs.] It was like, "No, no man's going to do this." I've always counted on a man before, and now I'm doing it myself, on my terms. That makes me feel really good.

What parent can forget the toddler's stubborn insistence—"I do it myself!"—and the elation when the shirt goes on or the blocks stay up? Independence is pleasing at any age, whether the attendant freedoms are as trivial as being able to tie one's own shoes, or as significant as choosing where to live and what kind of life will be led within those walls. Because women make less money and so have fewer choices, and because it's easier to raise children with a partner, they are generally less independent than men. Hitching one's wagon to a man-powered star is still the conventional scenario, and marriage institutionalizes and romanticizes this dependence.

But life on the lam has its distinct rewards, as Susannah discovered on her very first night in her own New York apartment. "I suddenly felt this unbelievable relief, physically, as though pounds were taken off my shoulders. I took a bath and nobody said I couldn't. I just felt like yelling, it was so great," she says with a grin. "I had so much energy that summer. I ran five miles a day, I could do anything. You know the best thing about being on your own?" she asks rhetorically. "When you're on your own, and the kid has diarrhea, you know you're the one who has to clean it up. But when you're living with somebody, you get pissed, you say, 'Whose turn is it?'" Susannah never missed Stan as a husband, and the only times she missed him as a father were at milestones like Eliza's first day of school. "That felt much lonelier," she says, "but everything else was a hundred times easier."

Susannah's basically positive, take-charge nature was a help, as was her conviction that "the big success was getting out." She recalls experiencing a grim moment of truth just after her divorce came through and before she had established herself in her film career.

I was standing in line and I realized that after I paid the baby-sitter I would have $80 left to my name. Tears just welled up and started flowing down my cheeks, and the thought that went through my head was, "I'm such a failure. I'm twenty-nine and a half and no one's taking care of me." And when *that* clicked—sometimes you have to say things to yourself to hear, "Boy, *that's* fucked"—I realized, "Wait a minute. I'm supposed to be so liberated, and actually deep down I feel like I'm a failure because I *have* to do this, as opposed to *wanting* to do this?" Just saying that changed the whole thing in my head.

The transition from dependence, whether real or imagined, to genuine self-reliance feels great. It also taps into a wellspring of ambition and ability.

They Know They're Competent

WENDY

I learned a lot about myself. I learned that I'm not a helpless female. I have a lot of confidence in myself.

For most women getting divorced ushers in new feelings of competence. The transformation was particularly sweet for wives who were belittled or undercut by their husbands, but every woman who discovered that she could indeed change the oil or manage her own money benefited by the realization. It's especially pleasing in view of the fact that incompetence in women is culturally sanctioned. When transsexual writer Jan Morris crossed the gender line she soon observed that "the more I was treated as a woman, the more woman I became. . . . If I was assumed to be incompetent at reversing cars, or opening bottles, oddly incompetent I found myself becoming. . . . I discovered that even now men prefer women to be less informed, less able, less talkative, and certainly less self-centered than they are themselves; so generally I obliged them."

Overturning the social order invites chaos and disorientation. "This is wrecking our lives! You've lost your mind!" husbands shout, and for a while it rings true. Ignorance and incompetence seem one and the same; the world seems full of alien transactions and mysterious machinery. But since a woman isn't expected to be able to fix the screen door, there's a bigger bonus in her eyes and others' when she pulls it off. Away from denigrating messages, faced daily with de facto evidence that they are coping, indeed thriving, women can't help but begin to believe in their own capabilities.

Beverley is no longer the dependent person she was when she married the handsome minister at her community church at age twenty. "Oh, that's something for Percy to do," she used to say during her marriage. "Oh, no, I can't do that." Now she can get a bookcase that arrives in a thousand pieces, "and sit down—I have my tools, my friends tease me about my tools—and put it together." She's learned how to set a budget too, and to explain it to others. That doesn't mean that Beverley liked acquiring these skills. "I thought it was wrong that I would *have* to do it, whether or not I wanted to," she says. "But now taking care of myself is part of life, there isn't anyone else out there to do it." Her self-reliance has made her more assertive too, which has made a difference at work: "I no longer allow people to do some of the things they did before that would bother me."

Winifred's divorce settlement came in the form of a lump sum, which she realized she had better invest wisely. It was alien terrain. "I didn't know how to talk to doctors, lawyers, real estate people, investors, anything," she admits. "I knew how to talk to gas station attendants, you know, like, 'Fill it up with unleaded, please?' I felt like there was this completely different language. I was this Annie Hall, zany and irrational." She did her homework, "consciously said, 'I've got to learn this,'" and now dispenses investment advice to friends.

> I even remember at one point I had gotten a CD at a bank and it had finished or something. I was inexperienced and the woman knew that, so she was trying to pull the wool over my eyes and give me a lower rate. And I said to her, literally, "Don't fuck with me." She blanched. I said, "You told me it would be this, and don't change your rule on me." She did it.

They're Happy

MARELLA
The most important lesson I learned from ending the marriage is that you should always feel happiness. That is your right.

What could be better than a fresh start in life, especially on one's own terms? Vulnerability can be demoralizing, and self-reliance is uplifting. It makes sense that as the misery of an unhappy marriage and the confusion that characterizes the period of transition recede, happiness takes their place. Peaceful homes, fulfilling careers, and new relationships all played their part in the overall sense of well-being reported by the women interviewed for this book. A 1982 survey of divorced people a year after the breakup found that 60 percent of the women were happier, compared with half the men, with the majority of women saying they now had more self-respect. By the five-year mark, according to Judith Wallerstein's study, two-thirds of the women were more satisfied with their lives, and five years after that, 80 percent said it had been the right thing to do. (The proportion of men who felt the same way held steady throughout at 50 percent.)

While she was married Keiko considered herself part of a specific segment of society: well-educated women who marry successful men and end up completely stifled. "They run the kids to school, they pick up the dry cleaning, they order their houses, and then, if there's time left, they work. But what they do in life is secondary to what their husbands do; they become facilitators, concierges." That description fit Keiko too, and though she misses the financial security of marriage, she's far happier out of it. As she explains it, "My life is much fuller. I see myself as being complete, in a way that I didn't see myself before. I'm taking more risks, not because I'm alone but because I'm no longer in this marriage." In sum, "I feel capable."

Those prosperous suburban wives no longer socialize with Keiko, and she knows why. "I'm really threatening," she says matter-of-factly. "They can't deal with the fact that I'm struggling, but dammit, I look good and I look happy! They're envious. It's power, personal power." These days, Keiko feels plenty powerful.

> I see these women in the grocery store, they're there with their kids, they look horrible, and I come in, even after a day of work when maybe I didn't earn anything, and I've got my child and I look good and I feel good. And I say to myself, "You stupid, stupid women. Why stay in a marriage if you're so unhappy, so unfulfilled?" It makes me feel like a tiger. I feel like saying, "RRRRRROOOOARR!"

Delores, the young woman of Cuban descent, has worked through fear, rejection by her family, and homelessness to a place of serene contentment. "My parents will never accept I'm divorced, so everything in life has a price," she says bluntly, "but I am who I am, and my son is proud of me." The fact that she's unattached is absolutely fine with her. "Marry again or don't marry, have a happy life," she advises. "If I meet someone, I ask, 'What would I be winning out of this marriage?' because I'm alone and I'm happy, honestly." Indeed, her face is shining with pride and satisfaction. "If I meet a man it's to continue being the same happy or get better, but not saying, like a lot of women, 'I just need somebody for my son,' or whatever. You don't need nobody. You just need yourself." Delores achieved this peace of mind at great personal cost, and it's hers for life.

A second marriage is one reason Helen, the woman from Tallahassee, is now so happy. "I've got a *wonderful* husband," she declares, beaming, but, like Delores, she wasn't interested in remarriage until it felt like she wasn't giving anything up. What she had learned by then, she says, is "that marriage was not a state of care. It was a state of choice." She tied the knot with Hank because "there was no sacrifice, it was like all he could do was enhance my life." She adds, "I genuinely like as well as love him, and I think he feels that way about me." About her first husband, on the other hand, Helen says, "I didn't like him, and I didn't like who I was either." She considers the old Helen and the present Helen to be two different people. The post-divorce woman is happy, "and I think she was happy because she learned who she was," says Helen. "Number one, she found out she wasn't stupid. Number two, she found out she could take care of herself."

SOPHIA
Divorce was not the end. It was more of a beginning for me.

For the women in this book, resolution and contentment concluded a harrowing and rewarding emotional journey. Divorce was the catalyst, perhaps not one they would have chosen, but worth it in the end. It's not easy to tune out our culture's characterization of a woman who goes after what she wants as selfish and immature. "The world doesn't like independent women," said ninety-two-year-old photographer Berenice Abbott. "Why, I don't know, but I don't care." Like Abbott, the women in this book have come to value themselves over obedience to an abstract and often punishing ideal. Without exception, they knew they wanted something better, and they finally believed that they deserved it. It feels great.

Their insistence on personal fulfillment and their refusal to equate it with selfishness benefited not only these women but their families as well. Research on the effect of divorce on children is wide-ranging, but one consensus stands out: their well-being is directly linked to the happiness and stability of the custodial parent.

CHAPTER 6

Kids and Moms

*T*here's one big plus to divorces that involve children: kids are incontrovertible evidence that the marriage was not a mistake. Everything else about the situation is wretched: kids grieve, their lives are reduced to a commodity to be sliced up like pizza, and their parents battle over the highest of stakes. Adding a surreal element to all the stress on the divorcing woman is that to protect the children who still are shared, she must continue a relationship with the very person from whom body and soul are disengaging.

Though full of remorse for the suffering it caused their families, especially in the immediate aftermath of the divorce, with only one exception the mothers interviewed for this book do not feel that the experience ultimately harmed their children. In fact, the majority feel their children actually benefited from the change, in ways both predictable and unexpected. Their experiences reflect what studies—including a twenty-year survey of twenty thousand families—have shown but what nevertheless is still not common knowledge: that divorce need not damage children. In fact they do well after divorce if they have the love and support of just one parent. If both parents maintain strong connections to the children, kids do even better.

How it would have comforted me to know this at the time. By far the worst part of my own divorce was agonizing over the effect it would have on our daughter and son, then seven and five. I remembered all too well the horrific moment two years earlier, when my husband blurted, "Maybe Mommy and Daddy are going to have to get a divorce." My daughter burst into tears and howled, "What's going to happen to *us*?" I wondered how I could possibly put my needs above my children's—it seemed a clear choice—and banished the prospect of divorce as too hideous to contemplate. As I later learned, the knee-jerk assumption that a divorcing mother's interests and those of her children cannot coincide is punishing and false, but I sure felt selfish at the time. During my interview with Olivia I confessed that I felt particularly horrible about having robbed my children of the sight of their parents strolling down the beach together thirty years hence. "Your daily life is more important than one walk down the beach," she chided me in response.

Another woman I interviewed, Laurie, confessed she had been bulldozed by guilt. "My ex-husband said, 'You're ruining my life.' My son Sean said, 'You're ruining my life,'" she recalls, still looking stricken. A friend admonished her, providing a much-needed reality check when she said, "Laurie, Sean's going to be ten years old. Next time you turn around, he's going to be leaving for college. I know you love him, but make a life for yourself. Don't stand there and end up with nothing." When her husband moved out, Laurie felt terrible about the great unhappiness she had caused him and her child, yet at the same time she felt "lighter than ever, and optimistic that I could have a happy life." Her optimism was borne out. A senior in college, Sean has a good relationship with his dad, an affectionate stepfather, and two half-brothers who worship him.

Asked what advice they would give to other divorcing mothers, each of these women urges that mothers acknowledge how disruptive divorce can be for children and work hard to prepare and protect their kids from the consequences. Laurie's advice is typical. "You really have to look into how old the children are, their emotional state at the time, how you're going to take care of them on your own. I'm not saying forget about divorce and stay married for the children, but it should be a very big factor in when and how you do it," she counsels. "Then," she adds with a big smile, "go on with your life. Give yourself a chance to be happy."

Children should indeed grow up knowing that they matter most. When their parents divorce they suffer disruption and often great loss, and that loss is permanent. But when mothers automatically and unquestioningly and endlessly sacrifice for their families, ultimately no one benefits. Children whose mothers are profoundly depressed, or whose parents are constantly fighting, thrive when those circumstances change. The gains for children in less dire situations are less tangible and the losses more confusing, but they too do just fine if parents remain committed to their welfare. In a lovely irony, my children now get far more of their father's and my undivided attention than they did when we were married. Though I could never have imagined it at the time, and though their grief is real and ongoing, my children are better off because their father and I are no longer together. So too, I believe, are the majority of other children of divorce and their new families, despite the logistical and financial hardships faced by the women who usually hold them together.

The Economic Plight of the Divorced Mother

Contradictory to all the family values rhetoric, Arlie Hochschild points out that "We do not have a family-friendly society." She calls it a "stalled revolution," explaining that "women have gone into the labor force, but not much else has changed to adapt to that new situation. We have not rewired the notion of manhood so that it makes sense to participate at home. Marriage then becomes the shock absorber of those strains."

Women struggle to fulfill old and new responsibilities, at considerable psychic cost. They often fail to give themselves credit for all they accomplish as economic providers, parents, and full-time managers. The only sanctioned role model for a woman who competes in the workplace while looking after her family is the nanny-and-briefcase Supermom, which is both costly and grueling. When their marriages fall apart under these strains, most women become single mothers, and single motherhood means financial hardship. (Though not all working mothers are single, all but a minuscule minority of single mothers in the United States work, which makes the terms nearly interchangeable.) More than 40 percent of divorced women are not awarded any child support, and only half of those who are receive the full amount from their ex-husbands.

Violet, who now earns a decent salary at a battered women's shelter, is one of the great majority of women interviewed for this book for whom the emotional difficulties of raising children alone paled besides the stress of trying to make ends meet. She's also among the majority of those whose ex-husband's standard of living improved post-divorce and who constantly confronted the galling reminder that her ex could provide things she could not afford. Violet faced dual stresses: poverty itself, and coping with the anger of kids too young to understand why their lifestyle has taken a dive. As the years passed, Violet's ex-husband saw less and less of the kids—"he used to drive past us as we walked down the street and not even honk, and the kids would cry and cry," she recalls— but he did take them over the occasional weekend. Sunday nights when they returned were always troubled. The little one would have tantrums, the two older ones would squabble, and then they'd turn on their mother: "'Dad has a pool table in his house. Dad has a freezer full of meat and look at our refrigerator. We don't have nothing. You're not doing right, Ma.'"

"I knew they were trying to vent, but it still hurt to hear your kids come down on you like that," admits Violet. "I did a lot of talking trying to steer them the right way. They're all in their twenties now, and they've all told me how they really appreciate how I kept them," she says philosophically. "I guess in the long run it's proven good." Part of what Violet was experiencing was a common phenomenon: children are often on their best behavior with the noncustodial parent and fall apart when they get home. Though this is aggravating for the custodial parent, it's comforting to realize that the good behavior is often motivated by fears of abandonment and that tantrums can be a sign that a child feels secure.

Violet didn't choose to be a single mother, but her situation has become an astonishingly common one. Given this trend, we should not be surprised that women who can afford to have children on their own are increasingly choosing to do so, despite the flak. The backlash against these mothers is summed up in an open letter to Madonna by *Newsweek* columnist Jonathan Alter, urging the pregnant pop star to do the right thing and marry the father. "Part of your message seems to be that you don't *need* men," he points out, observing that "this is now an extraordinarily common view, and it's substantiated by a long and sordid record of mistreatment of women by men. The problem is that it's also an extraordinary destructive view." Why? "A little quiz," posits Alter condescendingly. "The most accurate predictor for whether a child will drop out of school, face unemployment, and commit crime is what? Poverty? No. Race? No. Neighborhood? Nope. The answer is growing up in a single-parent family. It's the only killer correlation."

Actually, the real killer correlation is between single-parent homes and poverty, and the primary cause of that poverty is unpaid child support and the wage gap. What hasn't occurred to Alter is the possibility that Madonna, like Murphy Brown, is staying single because she can afford to. Unfortunately, women without celebrity bank accounts can't afford decent child care, they don't receive child support, and job training is scanty, so they are stuck at the bottom of the economic ladder. They can't afford dentists or good schools or even groceries. They are often depressed and angry and short on coping skills. Their kids grow up without male role models in landscapes devoid of opportunity. *That's* the problem, not marital status. "Research shows again and again that poverty and unemployment can more reliably predict who will marry, divorce, or commit and suffer domestic violence than can the best-tuned measure of values yet devised," writes Judith Stacey, a social scientist at University of California, Davis, pointing out that a secure income and a living wage are what genuinely shore up marriage.

Adding to the anxiety of divorced women is the fact that working full-time, a financial necessity, deprives them of time with their children. Buckets of propaganda reinforce the attendant guilt, but studies show that fathers, day care facilities, nannies, baby-sitters, kindergartens, and boarding schools all care for happy, well-adjusted children. The most recent and comprehensive study, conducted on 1,300 families over a five-year period by the National

Institute of Child Health and Development, found that even infant day care does not interfere with the mother-child bond. Any harm done to families, the study found, is caused by *anxiety* over child care and the needless stress it generates. Fifty years of studies show virtually no differences, by any measure of child development, between the children of working mothers and those of homemakers. When this becomes common knowledge, perhaps single mothers will stop beating themselves up and start agitating for nationally subsidized child care. As law and history professor Mary Frances Berry writes, "The issue of child care is really an issue of power, resources and control among adults; it is not a battle over who is more suited for care."

What Is a "Family"?

In recent years much has been made of the demise of the beleaguered American family. In fact the family is far from defunct, but as it struggles to accommodate the social revolution of the past thirty years, it is changing form. Because the roles of men and women at work and at home are in radical transition, the definition of "family" needs updating. The proportion of "traditional families" in the United States has declined by 35 percent since 1970. By the year 2000, nearly half the population will be part of a stepfamily. Sixty percent of American families are headed by a single parent, and more than half of those parents have never been married at all.

Washington's "pro-family" groups are busy portraying these statistics as a pernicious trend undermining women's most important role: staying home and bringing up babies. "The top priority this nation faces for the next generation," declared Alan Keyes, a former Reagan Administration official in a 1995 campaign speech, "is the restoration of moral and material foundations of the marriage-based, two-parent family, pure and simple." Conservatives hold up a simpler, safer era, in which kids walked home to Mom banging their lunchboxes on the picket fence, as the way it used to be and the ideal to which we must return. Even if we could turn back the clock, this scenario is not the way things used to be—nor is it ideal.

Debunking the Nuclear-Family Ideal

The postwar "golden era" of the nuclear family, with Dad the breadwinner and Mom the homemaker, was actually a blip in America's history that lasted for less than a decade. Not until the mid-nineteenth century did courts begin questioning fathers' absolute ownership of their children and identifying mothers as the principal caregivers and granting them custody. As discussed in Chapter Four, it was not until the Industrial Revolution that women became responsible for the unpaid work of child care, as production shifted out of the home. Postwar child care policies encouraged women to be homemakers, and those who did work turned largely to other women, whether family or friends, to watch their kids. Mothers who wanted day care had a very hard time finding it, and were guilt-ridden if they succeeded because of its association with poor and dysfunctional families. That stigma lingers to this day, as does the shortage of options for working parents.

Certainly more women than ever were able to stay at home with their children during those prosperous postwar years, but these mothers remained a privileged minority, and now represent only 3 percent of the population. Full-time homemaking was never an option for the majority of the working class, or for families of color. And a collective amnesia prevails about the dark side of those postwar years, which were characterized by soaring rates of juvenile delinquency, teenage pregnancy, and depression among tranquilizer-popping stay-at-home moms. Amnesia prevails too when it comes to the fact that many prominent family values spokesmen—including Bob Dole, Phil Gramm, and Newt Gingrich—are themselves on their second wives.

Romanticizing the traditional nuclear family is a mistake, because growing up with bio-mom and bio-dad is no guarantee of psychological well-being or financial security. What's more, that nostalgic and narrow-minded definition of family excludes not just the divorced, but also widows and widowers, adopted and foster children, and all those who love and are loved outside a legal contract. It sanctions job discrimination against parents who work, and who need all the help they can get. It ignores the fact that divorce simply brings to light problems that were already present (and that continue to seethe privately and damagingly in many intact families), and that divorce is very often the right decision for both the adults and the children involved. It denies the reality that many divorced parents continue to collaborate successfully in raising healthy children. It perpetuates the myth that divorced people do not honor or value marriage. It's elitist, anti-feminist, racist, and homophobic as well, because gay people are good parents too, as many studies evidence. For all its rhetoric, the family values agenda is profoundly anti-family at its core.

The Underpinnings of the Family Values Mania

Everyone seems to be "pro-family" these days: feminists, the religious right, gays claiming the right to marry and parent without prejudice. (After all, who would call themselves "anti-family"?) Everyone's jumping on the bandwagon because images of family breakdown are so prevalent and unnerving. Talk shows, sitcoms, made-for-TV movies, therapists' offices, and 12-Step programs are filled to overflowing with offspring raised in the dysfunctional variety: victims of incest, marital rape, paternal abandonment, beatings, even satanic ritual abuse. Home, in fact, is where women are most likely to be beaten or raped and children molested or killed, and not by strangers but by members of their own families. "At some deep, queasy Freudian level we all know this," notes Barbara Ehrenreich, going on to comment that "Americans act out their ambivalence towards the family without ever owning up to it."

The problems confronting the American family are cause for genuine concern, but the comforting reality is that true family values—the loving collaboration between family members who nurture and look out for each other—are not inherently dependent on Dad-as-protector. The current family values mania is part of a last-ditch effort to find stability in an era of shrinking economic and personal security, and to restore to men a clear-cut definition of masculinity. Male authority, already diminished in the workplace by growing intolerance of sex discrimination and harassment, is now waning on the home front. Seeing their power base erode, many men are exhorting a return to a family life that never was and in which their sovereignty over women and children was unquestioned. Many women also buy into the false security of the "good old days," but it's too late to stuff the genie back into the bottle or the woman back into the kitchen. Too much turf has already changed hands, and a more equal distribution of responsibilities within a family benefits all its members.

The Roots of the False Assumption that Divorce Devastates Children

A key element of the family values agenda is the assumption that divorce ruins children's lives, and it finds an easy foothold among guilt-ridden parents. The flames have been fanned by the media, which is quick to publicize findings that reinforce that assumption. Much hand-wringing followed the publication of Judith Wallerstein and Sandra Blakeslee's 1990 book *Second Chances*, a study of families ten years after divorce, that depicted children emotionally damaged for life, even though the study was limited in scope and had no control group. Let's face it: bulletins about the catastrophic effects of divorce sell a lot of soap. Another study, published in 1995, called "Growing Up with a Single Parent," concluded that children from single-parent homes fare less well at every stage of life. It's frequently quoted by conservatives to prove that the divorce is the problem and to justify the repeal of no-fault laws. That simplistic conclusion, however, is not the one arrived at by the authors of the study, Professors Sara McLanahan and Gary Sandefur. "People who get divorced probably should get divorced," says Sandefur. "The worst thing for kids is to be around a constant state of warfare." McLanahan calls the movement against no-fault "a bit of a sham. Just an easy fix that will appeal to voters."

If children of divorce suffer psychological problems, it's usually because of the troubled family life that preceded the breakup, not the divorce itself. Research seldom compares the effects of divorce to those of other major upheavals in family life, or takes into account other factors such as downward mobility, so negative stereotypes tend to be reinforced, as sociologist Terry Arendell notes. But they are based on myths and reinforced by guilt.

A Good Divorce Is Better for Children than a Bad Marriage

LORRAINE
I know what's good for me is good for my kids.

During the 1970s the conventional wisdom that unhappily married parents should hang in there "for the good of the children" yielded to a general acknowledgment that divorce is less stressful for families than life in a battle zone or an emotional freezer. Despite a recent, nostalgia-driven lament for the days when Mom and Dad toughed it out, there is no evidence that the good old days were better for children and considerable evidence to the contrary.

Comparing children raised in high-conflict families where conflict led to divorce with those whose high-conflict families remained intact, researchers found that children had similar problems. As summarized by Rosalind Barnett and Caryl Rivers in their study of two-income families, "The researchers said that if children emerged from the trauma of divorce with a strong relation with a parent they were *better off* than if they remained in an intact family in constant upheaval. They challenged the notion, that divorce caused permanent and long-term damage to children." Divorce releases many mothers from tension and depression. Happier, they are able to be more generous with their kids; calmer, they are more even-tempered; self-reliant, they provide better examples of how to make goals and live up to them.

Susannah's easy give-and-take with her second husband provides a wonderful model for Eliza, now college-age, and their twelve-year-old son. She says, "Any kid worth his salt would notice that there was no affection in the home beforehand, to say the least, or that their mother was resentful. Better you should say, 'I'm really happy with who I am now. I'm doing what I need to do for me.'" Susannah has complete faith that what's right for her is what's right for her child, and that Eliza benefited immediately from her mother's happier state. Two at the time of her parents' divorce, Eliza is a relaxed and flexible twenty-year-old who, says her mother, "comes to her own decisions about who her mom is and who her dad is."

HANNAH

If you're in a marriage that is really not healthy for you, and you're dying in it, it's killing your spirit or your mind, you're not protecting your children from anything by staying in it.

One reason that children of divorce are often independent thinkers is the example set by a strong and autonomous parent. "I don't mean to sound egotistical, but I think the kids really got the benefit of my being a single parent," says Marella, who had three children with her Nigerian husband and then twins out of wedlock, "because I was determined not to let anything stand in my way in terms of them getting an education, in terms of their manners, the way they interact with people." Had she stayed married, Marella feels she would have been a much more distracted parent, and that "I would have taken out the frustration and unhappiness of being cooped up with him on the kids. Now I'm happy."

She feels the divorce was hardest on her oldest child, seven at the time. "Sometimes he would ask me, 'Where's Daddy?' and that took a lot out of me," Marella admits. "That is the absolute worst." She did her best to reassure him that there were lots of people in the world with just a mommy, reminding him of all the aunties and uncles who loved him; much of Marella's extended family has also emigrated from Trinidad. Ultimately Marella feels that the divorce brought out the best in her and her children, "in terms of them knowing that life is a struggle sometimes but if you stay strong and focused, you can overcome that struggle," she says firmly. "I tell them that every day. They have all done well."

Theresa too was infinitely happier post-divorce, even though at first it didn't seem that way to Nora, the oldest of her three daughters. "I bet Daddy left because you were always yelling at him," said five-year-old Nora accusingly, soon after the separation; Theresa, a nurse, had always been the disciplinarian, the parent who dragged the kids away from the television or who got upset when her ex-husband, David, let them stay up late on school nights. But those issues resolved themselves over the course of a year or so, and Theresa believes that the divorce has definitely benefited the children. Short-tempered and self-centered, David never had much patience with his daughters. "Short-term they love seeing him, and then when he gets grumpy he goes home, which they really like. And *I'm* not as grumpy anymore, don't yell as much. So they have these two parents that are happier, and I think they sense that." Theresa's also very nice to her ex-husband in front of the kids, and says that for their sake she'd sacrifice a lot to keep it that way. As a single mother she feels much more able to handle the kids' affairs in their best interests. "I guess it's a control issue," she says frankly. "I feel I know what's best for them and I can provide that without interference."

NANCY

Be very honest with kids. They don't need to know a lot of details, but I'll say, "You know, Lauren, Mom and Dad don't get along. And I'm really mad at your daddy now, but it has nothing to do with you. He loves you completely and so do I, and that's what you really need to know." Keep telling them that, and somehow I think it sinks in.

Though acutely tuned to the ways in which their children have borne the brunt of divorce, the mothers in this study believe their children are better off now. If how their children are doing is any indicator that the divorce was not a mistake—and what better one could there be?—they're right. Think too about how many adult children of divorced parents say it vastly improved their home lives, or that they wished it had happened years earlier. They ought to know.

Bringing Children Safely Through Divorce

Some divorces do benefit one spouse at the expense of the rest of the family, which breeds an atmosphere of resentment and can result in ugly and ongoing conflict. But there are good and bad ways to divorce, and tried-and-true rules for bringing children through it safely and on to better lives. What makes a divorce agreement work for the children is not just who sleeps where, or how often people see each other, but an agreement that reflects a genuine commitment on the part of all the adults involved to put the children's needs first—*really* first.

The Ground Rules for Divorcing Parents

Younger children in particular usually prefer the known parameters of an unhappy home to the unknowns of divorce. The dismantling of the only way of life they know is terrifying. They feel very vulnerable, and they worry about parents who themselves seem newly so. Can Mom cope on her own? Where will Dad live? Who will take care of them? Signs of depression and anxiety commonly follow on the heels of the separation itself. On the other hand, children freed from conflict-filled situations or a disturbed or abusive parent may blossom immediately. Most children do cope and adjust successfully if the custodial parent is satisfied with his or her life and able to provide a stable and supportive home.

A parent has little control over some factors, such as the temperament of the child and how well the other parent will adjust to the arrangement. Especially during the tumultuous period immediately following the separation, even the most conscientious parent may be too busy keeping her head above water to give children the attention they need at this stressful time. Nevertheless, experts agree on two ground rules that minimize the negative impact of divorce on kids:

1. *Maintain the children's network of favorite family and friends.* Whether this means visiting grandparents, arranging play dates, or inviting favored aunts and uncles over, it's crucial that children feel that people who matter to them are still in place despite the schism at home. It's also a good idea to keep as many of the family routines—mealtimes, movie nights, holiday rituals—in place as possible. Because they make life feel more predictable, routines make children feel secure even in the face of upheaval.

2. *Support and cooperate with the other parent as much possible, and keep kids out of any conflicts that do arise.* This is especially difficult during the immediate aftermath of the separation when hostilities are at their peak. But if you and your ex-husband can manage it, and accommodate each other when necessary, the whole family will benefit enormously.

Children whose parents abide by these rules will get through their parents' divorces sadder and wiser but psychologically whole. "If you do these two things," says Andrew Cherlin, an expert on divorce at Johns Hopkins University, "and the kids were doing well beforehand—then, after a period of adjustment, they will do well again." Dr. Constance Ahrons, a specialist in children and divorce, is even less equivocal, stating, *"These two factors differentiate between the children who are and are not damaged by divorce."*

A few more pointers:

3. *Don't criticize Dad.* It's okay to vent, but to other adults, not to the kids. Children will figure out their father's strengths and weaknesses in their own time, without any help from Mom. Attempts to shape or deny the father-child attachment are unethical, and likely to backfire anyway. Besides, badmouthing Dad to the kids uses up a lot of energy better spent on more constructive activities.

4. *Don't put the child in the middle*; take questions or problems directly to the other parent. My kids usually beat me to it on this one. "You tell Daddy," says my daughter, thrusting the phone in my face and refusing to relay even the simplest message about a soccer game or orthodontist appointment.

5. *Never put children in a position of having to choose between parents.* When they are old enough, they'll make their choices clear enough, but until kids take that initiative their parents should present them with mutually agreed-upon decisions and arrangements.

"Everything that we did was geared to our daughter's well-being." Caroline

Caroline and Richard tried to minimize any damage to their daughter, Heather, by making their divorce truly child-centered; the papers actually stipulated that both parents do everything in their power to avoid using Heather in any disagreements that might arise. Obviously such an agreement is only as solid as the best intentions of both parents, but both Caroline and Richard have honored it over the years. The acrimonious divorce of Caroline's own parents colored her whole childhood, and she was determined not to put her daughter, two at the time, through the same ordeal when her eight-year marriage ended. Caroline's first move was to reassure the extended family that ties would be maintained. "Everybody thought I'd go back to England and take her with me; his parents were terrified. I said, 'Absolutely not.' I wasn't going to deprive her of him and deprive him of her." Heather spent holidays and time every summer at her paternal grandparents' house in the country, which worked well for everybody. Caroline and Richard also collaborated on a very flexible visitation arrangement, and Caroline footed the cost of college rather than wrangle with her ex or see her daughter denied. For Heather's sake, Caroline and Richard put themselves in their daughter's position, honestly and selflessly. It's not easy—it means picking one's battles, biting one's tongue, and keeping the inevitable irritations and conflicts to oneself—but it's the right move.

It's hard to manage this as smoothly as Caroline and Richard did. One strategy that can really help is to establish clear new boundaries for family relationships. Dr. Ahrons considers this essential, and suggests visualizing the binuclear family as a limited partnership whose partners follow clearly established rules. (Dr. Ahrons coined the term "binuclear" to describe a family that spans two households and continues to meet the needs of children.) Where tension is really high, she proposes a written contract detailing everything from whether a parent is allowed in the other parent's house to who's in charge of doctor's visits or setting bedtimes. It's demoralizing to have to fall back on a piece of paper, but it provides a course of action until heads have cooled and mutual accommodation is less remote a goal than peace in Palestine. "A good divorce does not require that parents share child-care responsibilities equally," explains Ahrons, but rather that "they share them clearly."

LAURIE
Try to explain to the children that it's not their fault, of course, that it has nothing to do with them. Keep the lines open, take the time to talk to them.

Any number of things—illnesses, remarriages, financial upsets, job transfers, children's schedules and wishes—can throw a crank into the arrangements, and they will. But if parents can step back, cool off, and look at each set of decisions as objectively as possible—from the child's point of view, that is—they'll be on the right track. At the suggestion of my ex-husband's attorney, we built a family counselor into the agreement to resolve child-related issues. Not only has she proved eminently sensible when we've consulted her, but just knowing someone's available to mediate is a comfort.

I'm still no good at negotiating with my ex-husband, and where the children are involved, reason has a way of taking flight.

When a Parent Doesn't Play by the Rules

The fact is that any agreement is maintained only by the goodwill of both parties. Some parents don't follow these guidelines, some because they can't and others because they won't. If this happens, by continuing to act responsibly, the other parent can minimize any damage to the children.

"Because he was so angry, he spent a lot of time badmouthing me." Rosemary

Rosemary, the air traffic controller in Reno, didn't realize ahead of time the extent to which her ex-husband, Gil, would vent his rage and frustration at the boys. Rosemary's boys were twelve and fourteen when they learned their father would be moving out, and she wishes she had prepared them more intensively, "perhaps by talking in general terms about, 'You know how some people don't get along . . . ?' for maybe weeks or a month ahead of time. My younger son reacted in the classic fashion of the child of divorce," she recalls with a shudder. "He jumped up when we told him, and started to cry, and said, 'It's my fault. I know it, I know it, it's my fault.' And ran to his room and locked the door and wouldn't talk to anybody, just cried and cried. It was awful." She and Gil explained that Gil would be a few miles away, that they'd spend weekends with him, and that he would be moving out on September first.

"We played at being really civilized," says Rosemary, but unfortunately Gil wasn't able to honor their agreement. Worse yet, he enlisted his sons in breaking it. With their help, he moved out of the house a week early while Rosemary was at work, on their wedding anniversary no less. "I thought it was really tacky," says Rosemary, and what made her even angrier was Gil later telling the kids he was depressed because their mother had made him move out on our wedding anniversary. "I thought that was really unkind, to try to punish the kids because he was angry at me."

Rosemary's sure that her children paid a price for Gil's anger at her having taken the initiative, and over his own unemployed status and consequent feeling of powerlessness. Gil was very critical of his ex-wife and hung posters in his apartment that said things like, "Men marry women because sheep can't cook." When the boys came back to their mother's house they would be full of rage, "and at that age they couldn't possibly understand that what he was doing was not about them and not about me, it was only about him," she points out. The older one really got it around age sixteen or seventeen, when, among other comments, he told his mother that he was glad his dad had a girlfriend because he could finally stop being so mad at Rosemary. The younger one, now eighteen, is still in the process.

Despite the anger and pain all around, Rosemary has no doubt whatsoever that the present arrangement is better for her sons. "Now they live in a stable environment with one happy person. After a while they understood that at their dad's the rules might be different but that was his choice, and when they were here this was their real life."

"He'd say, 'Be good, be good . . . and maybe your mother will let me stay.'" Lorraine

Lorraine's decision to wait until her children were eight and fourteen—"finally old enough to be okay if I had to spend a lot of energy elsewhere"—was probably a good thing in view of Jim's thoroughly inappropriate behavior both during and after the divorce. During the weeks before he moved out, as the children left for school with their mother he'd say, "Be good, be good. Your mother wants me out of the house. If you're really good and things are calm, maybe she'll let me stay." Though this manipulative behavior was almost certainly not deliberate on his part, Jim could not have found a more effective way to play on children's sublimi-nal fear that they had somehow caused the breach and should be able to heal it. As an elementary school teacher, Lorraine was acutely aware of the harm he was doing to them.

Jim still tells the children that Lorraine didn't give him a chance, and accuses her of having planned the breakup years in advance. Oblivious to their distress, he wanted his children to take sides. Not long ago he asked his daughter, Cheryl, to keep track on the calendar when Lorraine was visiting a boyfriend of whom he disap-proved. It didn't take Lorraine long to figure out what was going on, and to see that it put their daughter in an impossible position. Finally "I said to Jim, 'Look this is really uncomfortable for them, I'd like you to try not to do that,'" but then Jim scolded Cheryl for squealing. Their son has had some counseling for serious self-esteem problems. "He's come out of that tremendously in the last year," says his mother happily, but she's not sure counseling should get the credit. "I think it might be more the change in our family. He was put down by his dad all the time, very rarely supported." Recently his father attended a few sessions, which Lorraine thinks helped a lot. "Cody was able to tell both of us how he felt, not just about the divorce but what he'd lived with and what bothered him." Jim now sees the kids almost every week and occasionally takes his son fishing for the weekend. "I'm hoping their relationship is *better*, because I'm not in there, with the dynamics of us messing things up," says Lorraine. Because she herself is in good emotional shape and able to deal with her children's needs appropriately, they have all weathered the divorce well.

It is reassuring to know that kids are astonishingly resilient, though no conscientious parents want to put their children to the test. When Amanda, then in architecture school in Boston, moved out with their toddler son, her husband Anton's behavior was truly horrendous: he was incredibly abusive to her in front of Evan, reported her for child abuse, even pulled a gun on her. Amanda nevertheless made extraordinary efforts to support Anton's relationship with his son, but was ultimately forced to obtain an order of protection and supervised visitation. Sometimes Anton shows and sometimes he doesn't, and Evan worries that he'll hurt Amanda, "which is a tremendous burden, so much responsibility for a little boy," says his mother sadly. Nevertheless Evan is thriving. Now a sociable ten-year-old, he tells his mother what's bothering him, has no interest in seeing a therapist, is doing fine in school, and is deeply attached to his stepfather of two years.

The Reconfigured Family

WENDY

We wanted to keep this family but we didn't want to keep this marriage, I really think that's what it was all about.

Children of divorce still belong to families, "binuclear" or otherwise. When both parents are actively involved in their children's lives on a regular basis, it's not just the children who benefit. One factor consistently emerged during the course of Ahrons's research: fathers who stayed involved with their children had better relationships with their ex-wives than those who made themselves scarce. "One tightly-held secret," she writes, "is that out of many bad marriages come good divorces." Also, the more time passes, the less bearing a father's relationship with his ex-wife has on his relationship with his child, which assumes an independent life of its own.

Parenting Arrangements That Evolve Over Time

The active involvement of both parents usually—though not exclusively—occurs under the umbrella of joint custody, described in detail in Chapter Three. As with other aspects of the divorce agreement, the arrangement works as well as the best intentions of both parents. What's written in the divorce papers, though, tends to diminish in importance as time goes by and the demands of real life impose themselves. Ideally parents relax, regain trust in each other, and come to terms with what they can and cannot expect of their co-parenting arrangement. Family life is anything but static, so when parents can pull this off—and more often than not they do manage most of these transitions—everyone benefits. More important than the logistics of the arrangement at any given point is flexibility in trying to accommodate the needs of different family members as they evolve.

"I made it easy for him to change the schedule." Dina

Because Dina, an immigration attorney, was the initiator of her divorce, her husband Alan took a hard line in their negotiations: he refused to give Dina custody and insisted that she leave their house in downtown Sacramento. Agreeing to an arrangement by which her sons, then ages three and seven, alternated weeks with each parent is the only concession Dina regrets. She wouldn't agree to joint custody again, even if the alternative was a messy court battle. "I still think that was the source of a lot of problems my second son has. He was really young. And I hated being away from the kids a whole week." She believes that the older son "became the mother for the time they were at their dad's, doing everything for him." As a consequence the younger son still has separation anxiety. Nine years have gone by but "family is everything to him. He doesn't really connect that well to other people."

Fortunately the arrangement lasted for only two years. Dina gives her ex-husband credit for really giving it his best. "He really did try and play Mr. Mom. He made the birthday parties, he wanted to write a book on single fathers and how you do it." One of the reasons he didn't want the divorce, Dina surmises, "was that he thought he wouldn't be part of their life, from what he had seen from other divorces, and he wanted to be a big part of their life." But it was too hard. Alan went bankrupt, had to work long hours, and realized he couldn't handle it. He switched to having the kids for four days and a long weekend every two weeks, and then to a weekend every other week. "I made it easy for him to change the schedule," explains Dina. "I was always ready and willing to take the kids when he couldn't get home or this and that, and I finally said, 'Why don't I just keep them?' and he said 'Fine.'"

Dina was even more flexible, taking the generous and unusual step of moving closer to where her ex and his second wife lived, which made it easy for the kids to visit any time. Alan wasn't happy about it at the time—"I don't want my wife to bump into you at the grocery store," he protested, "I don't want you in my life anymore, it still hurts"—but it's worked out well. "Before, I decided where they went to school, camp, this or that," says Dina. "We make decisions more jointly now. The difference is that I ask him; it's not like he goes and does the research." Dina's arrangement exemplifies the situation in most American families: the father has more rights than responsibilities with regard to children, and the mother has more responsibilities than rights.

Dina has also looked the other way about Alan's default on his $400-a-month child support payments. He had stopped paying after four months, when Dina started dating someone else, and then ran into serious financial problems. Though Alan got back on his feet, Dina really didn't want to take him to court—"I'm basically a nonconfrontational person, and I decided that would hurt the kids"—so they worked out a deal that Alan would put the money he owed into an account for college. "I said, 'Fine,'" says

Dina, "but he hasn't done it." Because of the kids, though, she tries not to make money an issue. "I really have tried not to talk badly about their dad, much to my new husband's dismay. He would like them to know more, that Alan doesn't pay for this or that. But it would make me look bad, and if I even hint at something negative, they bristle."

The Pros and Cons of Shared Parenting

Mothering is hard even when you're not alone and even when you let go of any notion of achieving perfection, like television's Roseanne, who says, "If it's five o'clock and the kids are still alive—hey, I've done my job." A study of eighty divorced mothers in California found that those with the least contact with their ex-spouses were the most "angry, bitter, and resentful about the absence of a father from their children's lives—emotions that made it even harder for them to cope with the children's own feelings about paternal neglect." Understandably, bad feelings were aggravated when financial support was lacking. Sharing the job is ideal, but it can be a complex dance, requiring flexibility, generosity, and maturity.

A Child or Parent Can Become Overly Attached

Keiko is confident that divorce has made her a better mother to her four-year-old son, Asher, though she's not sure the same is true of Asher's father, Rick. "I'm happier than I ever was," she explains. "My son is getting the best that I can offer him now." What does concern Keiko is what she perceives as her ex-husband's excessive attachment to his only child. Rick has not sought out any other partners in the intervening two years, and "I think he's really having his emotional needs met through this relationship." Keiko adores her son, but she doesn't depend on him or smother him the way Rick does, and perhaps as a result, Asher's become very clingy. Also, rather than leave her child at home Keiko takes the boy everywhere with her, although she suspects that she should make sure Asher spends more time with other children. "But I'm trying to have a social life too, and this is the way that it's possible," she explains frankly.

The Back-and-Forth Takes a Toll on Mom

Caroline is very frank about both the rewards of two-family life and the toll it has taken on her and her daughter Heather, now twenty-three. She no longer considers herself a single mother, but while she was, it was a considerable source of pride. "'Yes, I'm a single mum.' I thought that was the bees' knees, that somehow or other I was coping and managing. I felt it was an achievement." She adds that in her relationship with her daughter, "there were times when I knew I could be tremendously overbearing, and at least we knew that over the weekend we'd get a break from each other." Caroline thinks all the moving around—"she has a room in her father's house, a room in my house, she had a room in her grandparents' house"—has not been good for Heather. Even though the perception at the time was, 'Oh it's fine, it's absolutely great,' I think that disruption ultimately takes its toll. It's emotionally upsetting, and makes it difficult to be organized."

The back-and-forth also takes an emotional toll on mothers.

Sometimes I didn't have her for Christmas, or I didn't have her for Thanksgiving, but that was my price to pay. I felt that I'd made the decision, and why should I mess up her enjoyment? Let her be with her grandparents and her father if it's what she wants. Since I got married again, it's been pretty much that she spends Thanksgiving with her grandparents and Christmas with me and my husband. Not all time. But that's the price you pay, and you can't divide their loyalties. You cannot. It's the same as using them for ammunition. And it's heartbreaking sometimes, especially at Christmas when you're with other people who have their children, it is *tough*. But, I say, "Hey, I'll have her another time, and as long as she's happy . . . "

Children Get More of Their Parents' Attention

I'm well acquainted with the tradeoff Caroline speaks of. The price I paid for my divorce is the sheer misery of having less time with my children, at having only every other Christmas and half of a vacation, making it difficult to plan a trip. Sometimes it's heaven to head off for a childless weekend, but other times guilt claws at me and I tell myself I've sold them up the river to obtain my freedom. Sometimes that seems like the simple truth, and sometimes it seems like pure masochism. For my part, I now cherish my time with them far more consciously. Instead of offering my standard refrain—"just let me finish this paragraph"—I now abandon my keyboard at three o'clock and swoop down on them hungrily. I also try to limit my nights out to the ones they spend at their father's. Some of this is because there's no man in the domestic picture, and some a function of the kids getting older, but I love the way the three of us are our own little gang. After all, soon they'll be teenagers and won't want to be caught dead with me.

One of the ironies of shared custody is that some fathers become far more dutiful dads than they had ever been when living under the same roof. I expected my ex-husband to weary of lunchbox duty in a matter of months, but he has come through admirably. When we were married, my husband competed with me by making himself unavailable for kid duty, taking them out only if I begged; now it's a competition to see who can log the most park time. Sometimes it bugs me, but I'm fortunate to share the awesome responsibility of raising my son and daughter with the only person who loves them as much as I do. And they benefit from the fact that neither parent now takes for granted the time that once seemed so endless.

So how are they? They seem fine, though I suppose the jury's out until they're parents themselves. Teachers and friends confirm my sense that both are happy and well-adjusted. If something's bothering them they usually tell us, and if not, we try and figure out what it is and get them to talk about it. I try to remember that their loss far exceeds mine, and I know they live with the grief of having the family they crave torn away from them. I also believe that this arrangement is fundamentally better, because children do as we do, not as we say. They see me in a healthy relationship with a man who delights in my competence, and they don't feel responsible for my emotional welfare. They also have an adoring, happily remarried father who remains at the center of their lives.

The Pros and Cons of Being the Primary Caregiver

The dynamics are quite different when a child spends most or all of the time with one parent. Sometimes this is dictated by a sole-custody agreement; in other cases it's simply what makes the most sense, especially when the father in question had little to do with day-to-day child care. Of the mothers in this study even those who sued for sole custody did not do so with the intention of separating father and child (although because of her ex-husband's threats of violence, Amanda was eventually forced to do just that). Even had they wished it were not so, a father's access to his children can be restricted only by a court ruling based on strictly defined evidence; the preferences of mothers and children and the father's child support record are immaterial. Because they were keenly aware of the advantages of shared parenting, going it alone was not the scenario most of these mothers would have chosen, though they cherish the closeness it has fostered.

The Primary Parent Takes the Heat

Divorce may require that mothers and children redefine certain aspects of their relationship. It's often difficult for mothers to assume the role of disciplinarian, a difficulty that can be compounded by guilt and reluctance to compare unfavorably with the other parent. Amy misses having another authority figure around when her headstrong five-year-old disobeys her. "She'll say, 'I hate it that you yell,' and I'll say, 'I hate it that you don't listen,'" Amy says resignedly. Martin, on the other hand, comes off smelling like a rose. "He is really quite wonderful with her," says Amy.

> One night she said, "Mommy, I like Daddy more than I like you." I said, "That's okay, honey, I understand." She said, "But is it okay to like one person in your family more than somebody else?" And I said, "Absolutely." I thought I handled it beautifully, and my child trusts me so much that she can say this, but inside it made me feel like shit.

Amy figures that any kid is likely to have more anger for the person the child lives with, "because Daddy comes and he goes." Ginny voices the same complaint: that her kids see being with Dad as "the new ice cream flavor. He has a trampoline, they're going to the movies, they're playing golf, and with me it's homework. We do fun things too, but it's still the heavy stuff." On the other hand, she says, "I was always the heavy." Underlying their resentment is the fact that they are among the many women whose altered circumstances require that they work longer hours than before. Amy manages her family's museum-supply business near her Oakland home, but Martin doesn't have to work. "He's always there for her, he picks her up, he cooks for her in a way that I would never do. She said to him while I was away these past ten days, 'You know, I eat much better with you,'" a comment Martin lost no time in passing along. "He never defends me," says Amy wearily. Though he will deal with issues if she makes her expectations clear, Martin remains basically unsupportive. "In a way this is actually saddest for me," she admits. "Martin lived like a single man when he was married to me, and he's still the same." She does see that two of them work hard to show their daughter Cara "that each of us is enough, because I think we feel that way."

The Responsibility Can Be Overwhelming

Virginia and Christopher have joint legal custody of their son, with primary physical custody awarded to her. Although Virginia feels better off than when she was married—"I see it as having one less child to support," she explains—she has a hard time making ends meet, especially with Ian now in private school.

After a long period of feeling liberated by being a single mother, Virginia no longer feels that way. "I'm tired of it, tired of all the responsibility all the time. I think to be this sort of overwhelmed with responsibility is ultimately not good for the child either," she adds. And while she's gotten rather comfortable with being able to do things how and when she wants to, in the long run she's not sure it's all that good for her either. For one thing, it's isolating, because, as Virginia points out, "you have to stay home with your kid, you have to make him dinner every night, you have to put him to bed every night, no matter what." And because Ian is an only child, Virginia also worries that the relationship is too intense, "too much, too much for me, too much for him." Now eleven, Ian has remarked a couple of times that he doesn't blame her for separating from his dad. "He's an anxious kid anyway, so the anxiety level stays high," comments his mother. "On the other hand, he has a wit and a perspicacity about human motivations and human behavior that surpasses almost everybody I know, child or grown-up." This emotional maturity, a significant consolation to Ian's mother, is a hallmark of children whose parents have divorced.

Yet for plenty of mothers the logistical transition to single motherhood wasn't all that big a deal, because they were already doing the lion's share of the child care, and when the myth of shared domestic responsibility was exposed as just that, it was a relief. Karen, the sculptor, says simply that when it came to caring for her son Ivan, "I did *everything*. . . . Neil sort of treated fatherhood like a guest-artist position. I was responsible for the weaving of the whole fabric, and he'd come along and sort of embroider a little something on it." Although eleven-year-old Ivan sees his father for dinner several times a week, Karen spent a fortune to obtain sole custody and insists that her son sleep in the same place every night, which means she has to be home every night at eight except on every other weekend. Having raised a much older child from her first marriage alone, Karen did not welcome being a single mother. "It's made my life much more restricted than it ever was when Neil was at home, and sometimes it feels like a tremendous sacrifice," she acknowledges matter-of-factly. "But it's been something I just decided that I was willing to do for the few years that it was going to be necessary."

In the wake of the divorce Neil has taken responsibility for himself and for his child in ways that never would have happened were he and Ivan's mother still married. For him, as for many divorced dads, it has meant being alone with his child far more, and far more regularly, than had ever been the case before. For noncustodial parents who can approach the situation in the right spirit, divorce provides a valuable opportunity for the parent-child relationship to be reevaluated and improved.

The Damage a Negligent Parent Can Do Is Minimized

Except for the financial hardship of raising her son on her own, Phoebe, the women's clinic administrator, says she wouldn't have traded her situation with anyone "for one second. My son and I are very, very close, very protective of each other. It's just wonderful. I've heard so many people say, 'Gee, I wish I had the relationship with my child that you have with yours.' My son will talk to me about anything and everything." She feels that staying in the marriage would actually have harmed the boy, because even when they lived under the same roof Howard was either disinterested in the baby or mean to him. "Brian had been my child from day one, because his father had chosen never to be there. The only time that he would play with our son was to start him crying. Then he'd laugh and walk on," Phoebe recalls, less angry than mystified. Though Howard had sued her for custody—a ploy to intimidate her—and was entitled to weekly visitation, "he never once came to pick up the child. Once in a blue moon, once a year, he'll call up and say, 'Hi, I'm Dad. Happy Birthday.'" Sad as that may seem, Phoebe is convinced that it was far better for Brian to be abandoned than abused.

Steve and Yolanda's intact family was far from nurturing to their third child, Will, who was born with serious learning disabilities. His much older siblings were impatient with him, and Steve was unable to accept the fact that there was something really wrong with *his* son; "he felt Will should just fucking shape up," explains Yolanda, the public relations director, about her domineering husband. When she finally saw how damaging this was, she left. Away from what she calls that "cruel, cruel environment," Will has come a long way, and so has Steve. Everyone's life improved when Steve remarried, to a woman who genuinely loves Will and acts as the disciplinarian when he's at his father's.

Mothers Without Custody

The other side of the picture, though a tiny minority, is represented by mothers who have given sole custody of the children to their ex-husbands. Certainly fathers are as capable as mothers of raising children; the myth that only Mom is up to the job damages families in a number of ways. Mothers who work full-time feel guilty, and fathers who stay home face the flip side of the same prejudice: that parenting is not a real job, and certainly not one fit for a real man. This attitude further feeds the stigma faced by mothers who give up custody, which is one reason they are so few in number.

Only two mothers in this study gave up custody, and as it happens, both suffered from mental illnesses that impaired their parenting abilities, as both women recognize. Illness, however, is only one of many legitimate reasons for giving up custody, whether temporarily or permanently. A mother may decide to go back to school and find the demands of job, housework, and children too much to handle. A job opportunity may mean relocating, and uprooting the children may not be in their best interests. Older children may refuse to make such a move. The children may need the companionship of the same-sex parent for a period of time, or they may not get along with a stepparent or stepsiblings, in which case life in the alternate household may be less conflict-ridden. Parents and children may benefit from time apart, especially during stormy adolescent years.

Irene abandoned her three children when she fled her repressive born-again husband, literally closing her eyes, pointing to a spot on a map, and ending up in Pierre, South Dakota. After three months of pain and soul-searching—"God, I ached for my children. I ached for them every day"—Irene returned home so that she wouldn't lose custody of her children. However, when she returned, and moved into a women's shelter, Irene learned that her husband had already filed for sole custody. She had spent time in a mental institution, and her husband threatened, "If you try to take the kids, I'll have you committed." Irene's bid for custody was unsuccessful, and she thinks that if it weren't for her lawyer's doggedness she wouldn't even have visitation rights.

Currently homeless and in school, Irene nevertheless pays $25 a month in child support of her own volition "because it balances the power," and sees the children when she can afford the plane fare. "I wish I had been stronger, to be able to take the kids and walk out the door," she says sadly. Though she isn't crazy about the religious extremism of her children's upbringing (their father burned all Irene's volumes of Shakespeare and destroyed her classical music CDs when she left because those works weren't Christian), she realizes the kids have benefited from an environment more stable than any she would have been able to provide.

Gloria also gave up custody, because she knows she's not Supermom. Hospitalized once for bipolar mood disorder, in the wake of her divorce Gloria found the demands of an eight-year-old from her first marriage, a toddler, and a baby too much to handle. Her older daughter still lives with her, but for the present she and Lee have agreed that the two younger children should stay with their dad.

Gloria sees her son and daughter every other weekend, from Friday until Sunday. She and Lee meet halfway, with each driving about forty-five minutes. "It's so painful, and so weird and draining and unsettling," she says with a deep sigh. "I can get real neurotic about it all, the mothering thing, the guilt. But I find I do better if I keep it to just a couple of days, so that's the way it is now." For a while Gloria felt absolutely horrible. Now, she says, "I'm pretty settled in my mind that I'm not ashamed like I was before." One thing she's learned to say is "'We share custody,' not 'My husband has custody and I visit.' And don't even bring it up, unless you've known the person for a while."

Gloria weeps as she describes the arrangement, but nevertheless firmly maintains that she's glad she did it that way. "With him everything's always the same, so I'm really glad that they are with the parent who is the rock during this formative time of their lives. I think that's important for kids, and I'm glad that I could see that that maybe wasn't my strong point, and it was his." Few divorced parents can put aside guilt or competitiveness to see their children's needs so clearly. It's a shame that the stigma is so strong that women in Gloria's position feel shamed by their selfless decisions.

When the Children Are Older

Clearly the older children are, the better they can evaluate their parents' situations, see their parents' needs separately from their own, and look after their own interests in the aftermath of divorce. Consequently they tend to be more open-minded and less fearful of the transition.

For Tara and Hal, telling their twelve- and fourteen-year-old children that they were getting a divorce was horrible, but they both worked hard to keep lines of communication open. Trying to figure out the logistics was a tense and uncertain business. For example, they decided to have dinner together one night a week, an idea quickly discarded when the first such gathering proved horribly strained. Still, as Tara, the massage therapist in Boulder, acknowledged, it really helped to have children old enough and articulate enough to say what they did and didn't need. Both of her children have spoken up about the original custody arrangement. Arrived at in mediation, it called for the children to spend three nights a week at their mother's and alternate weekends with their father, but soon after that one disastrous dinner Scott said, "Look,

Mommy, I'm fourteen years old. What I need is to separate from you and Daddy. I need a life outside the home. If you think I'm going to spend two days a week with you and two days a week with Daddy and one day with you together, you're crazy!"

Scott was also very clear about the fact that "he did not want to become my man. He said that," says Tara, still impressed by her son's maturity and forthrightness. It can be tempting to count on an older child to assume major responsibility for the household, or on an older son to become the "man of the house" in his father's absence, but such a burden can overwhelm the child, particularly if it is long-term. Luckily Scott was wise enough to nip this possibility in the bud.

More recently the schedule changed when Tara's adolescent daughter Sally decided she needed to be with her mother more; besides, she hated the school-night commute to her dad's. "That's hard for Hal. He misses her, he'd like to have her more often," Tara acknowledges. Fifteen-year-old Sally feels badly too, but this is the arrangement that's best for her right now. Tara thinks her relationship with her daughter has vastly improved since the separation, because Sally can no longer play her parents off against each other in the same way.

Tara's family has already coped with a common problem confronting adult children whose parents divorce: a parent may reverse roles and depend on the child for emotional support, as Esther found. She was delighted when her husband promptly remarried, because in the interim he'd been crying on his grown children's shoulders and really giving them a hard time. Even if they worry about an older parent's emotional upset or ability to cope on his or her own, adult children also benefit from having parents who are happy apart instead of miserable together.

OLIVIA

I think it's terribly important to stay in touch with your ex-husband, even with grown children.

Olivia's children, the youngest of whom was in college when she broke the news that she was leaving Duncan, all knew there had been a lot of tension at home. Of course it still came as a shock, but they were able to be there for their mother. One daughter said honestly, "Mom, 90 percent of me says this is just grand that you're getting a divorce. The other 10 percent says, 'But it's my mommy and my daddy.'" Shortly afterward, Olivia's son told her that it had made him and his brother "really reexamine ourselves as husbands and men." Olivia starts crying when she thinks about "what it's done to the children in terms of not having a family to come home to. That is tough." Christmas is the worst. Nevertheless she is delighted to be divorced, and clear that putting her interests first after thirty-five years was the right thing to do. "It's like when I moved away. I had to decide that I was living for *me*, not for my kids or for my life with my kids, because my life is every hour of every day."

Divorce asks the near-impossible of parents: to let go of some aspects of their partnership and to sustain others. As the saying goes, you divorce your husband, not your children, and they link you forever. Wendy, the cookbook writer, remembers reading a novel in which the female characters, all divorced, referred to their ex-spouses as their "biological husbands" because they would always be biologically connected through their children. "You cannot escape that," Wendy points out, "and the one thing I have done right is maintain a good relationship with Frank. We have celebrated our children's birthdays together, their graduations from law school together, the birth of our grandchildren together, and that's been the best thing that I've done." The fact that her children were well out of the nest didn't make this less of a priority for Wendy. Quitting the wife job is an option, but the mom job never ends.

Deadbeat Dads

Everyone benefits if loving fathers stay involved with their kids. Fathers reap the rewards of parenthood, children retain their most important male role model, and mothers have co-parenting and financial support. Statistics bear out the logical assumption that the more involved a father is with his children, the more likely he is to pay child support, since the outlay is likely to be motivated by love and concern for the children.

Yet which comes first, paying the support or maintaining the relationship, is not all that clear. In my interviews, many women reported that their ex-husbands paid not a penny of child support yet remained actively involved in their children's lives. Fifty-six percent of the fathers paid support regularly (though it was often a minimal amount) and were active parents; 8 percent paid nothing and had dropped out of their children's lives; and 36 percent saw their children regularly although they paid no child support. Economic circumstances may prevent some fathers from paying support, and it doesn't necessarily mean abdicating all the emotional responsibilities of parenting as well.

SEPARATING A FATHER'S EMOTIONAL CONTRIBUTION FROM HIS FINANCIAL ONE

Even if withholding visitation to get errant ex-husbands to pay up were legal, which it isn't, to the women I interviewed such behavior would have been morally unacceptable. Many cited reluctance to deprive their children of fathering as the main reason they postponed divorce for as long as they did. They bent over backwards to facilitate contact, adjusting vacation schedules, allowing visitation in their own homes, and buying plane tickets to help kids and fathers stay close. They didn't want their kids to suffer, and they wanted the children to know their fathers, warts and all, rather than live with fantasy versions and grow up with holes in their hearts.

Their instincts are correct: children who enjoy good relationships with their fathers are far more likely to see the divorce in a positive light (though a good relationship with a stepfather can fill the need, especially if children's pre-divorce relations with their fathers weren't good). The California Children of Divorce project found that good father-child relations appear linked to high self-esteem and the absence of depression, while irregular contact with the father correlates with feelings of rejection and a poor self-image. Where disruption occurred in what had previously been a loving relationship, children had a particularly hard time coming to terms with their loss.

"Can they eat a pair of sneakers? What is he, crazy?" Marella

Marella's husband Reg was one of the fathers who paid no support but still saw his three children regularly. Her family thought Marella was crazy not to go after her husband for the support he owed, and, she admits, "I cannot begin to tell you how many times I stayed without lunch so my kids could have dinner." A strong-minded Trinidadian, she was nevertheless determined not to go on welfare, or to take Reg to court so that a judge, a stranger, would tell him what he had to do. Also, she adds with a laugh, "To be quite honest, I felt that I was strong, and I wanted to prove it." Marella never stopped the kids from talking to Reg, "from knowing that this is their father, from going there." Awarded visitation every other weekend, Reg instead came by whenever he wanted.

Although she never tried to do so with her ex-husband, Marella has impressed a strong sense of responsibility upon her children. She remembers her older son saying, "Mother, why didn't you take him to court?" and replying that "I don't think a man should tell another man how to care of his kids." Marella does think the three children perceive their parents differently as a result of the divorce. "I think they see me as being very strong. And I think they see him as being very casual about the fact that he didn't give them any support financially." When the kids asked their father for money he'd offer to buy them school clothes.

> My aunt would say, "Can they eat a pair of sneakers? What is he, crazy?" He would say to my older son, "I'm putting money away for you for college," and that wasn't true either, because I paid for my son's college. So you can imagine the struggle and the constant battle that I undergo, almost every day, trying to stay alive.

"*I never said a bad word about him.*" Tory

Tory's husband also took no financial responsibility for his son, who was six weeks old when Tory walked out. He also saw the boy only very erratically, but Tory, now a well-paid human resources coordinator but hard up at the time, was never critical. "I see so many divorces where the mother tries to shape the child's vision of his father, and it's a bad situation," she says. "I'm real proud that I didn't tarnish my son's ability to have a relationship with his father." (Her ex-husband Mickey was clueless as to how unethical his behavior was and the hardship it caused his ex-wife. When he materialized six years later and suggested to Tory that they start seeing each other, Tory's response was to point out that since she wouldn't consider dating someone who had deserted his son, "why would I date the person who did it to *mine?*")

Tory's decision to put the father-child relationship ahead of any personal or financial failings on the part of her ex-husband was wise and generous. It is also typical of every single mother I interviewed, despite the characterization of divorcing moms as selfish and shortsighted. These mothers reasoned, correctly, that over time kids would come to their own conclusions about their fathers' financial and personal reliability.

"At first I tended to overcompensate." Beverley

After her ex-husband Percy made a halfhearted bid for custody when Beverley and their son Jermaine moved back to Roanoke, Virginia, after two years out of state, Beverley's husband Percy remarried and moved to Atlanta. Jermaine hasn't seen or talked to his father since, not even on his birthday. "That upsets me," says Beverley, "because my son had nothing to do with anything of this. I've never given him or his wife a reason to not contact my son, ever." At first she tried to cover for Percy's absence, indulging her son and taking time off from work when she couldn't really afford to. Now she's learned to tell him when she needs time to herself. "I'll set a timer and when the timer goes off then he gets my attention."

Another overcompensation on Beverley's part was covering up for her ex-husband's failings.

> At one time I was doing this thing of, "He's a wonderful guy, he'll contact you, don't worry about it," and it was actually one of his relatives who told me to stop doing that. She said, "You need to be honest with Jermaine. You don't need to make a habit of putting his dad down, but if he doesn't call, you don't need to say, 'Well, he thought that you were asleep.' He needs to grow up with a realistic portrayal of how both of you are." That was very good advice.

"I never wanted to be a single mother, and I don't like it," admits Beverley, "but there are some things between my son and me that are very good." One benefit of Percy's remarriage is that his second wife now sends Beverley the child support Percy owes. (His parents had stopped paying the $150 a month when Beverley moved away, and as a minister Percy gets no paycheck, but Beverley got some of the back support by intercepting his tax refunds.)

"My biggest mistake was not saying, 'I don't want child support.'" Phyllis

Sometimes money is less important than autonomy. Phyllis's ex-husband paid $65 a week for the two children, $15 of which was called alimony, until the kids were eighteen, though, as in Susannah's case, "if the kid needed sneakers when he was with them, I paid him back." Though not wealthy, Phyllis, a court stenographer, worked full-time after moving out of her mother's and has a small independent income. Ironically, Phyllis would rather have done without. "My biggest mistake, in a way, was not saying, 'I don't want alimony and I don't want child support. Go fuck yourself,'" she says bluntly. "If I had just cut him off, I wouldn't have been in this web for so long." Paying support "allowed him to still feel like he was the father." An abusive man with a violent temper, Phyllis's ex-husband had a negative influence on his children's lives; when the younger son attempted suicide at nineteen, it took his father five days to get to the hospital. Phyllis's husband used child support to maintain an element of control over his ex-wife and children, monitoring what they spent it on and approving or disapproving.

Although in this study there was no correlation between husbands who handled the family finances and ex-husbands who evaded their support payments, most husbands, mine included, resisted the notion of handing money over to their ex-wives and losing control over how it would be spent. Consciously or not, many men wanted to punish their ex-wives, and the fact that their children were punished as well held no sway. Laurie's ex-husband paid $50 a week until Sean turned eighteen. "He used to ask my son what I was doing with all the money he was giving me," recalls Laurie, with a snort of laughter. "Whoopin' it up!"

UNCOLLECTED CHILD SUPPORT—THE DISMAL FIGURES

As amply documented, unpaid child support is a prime factor in the impoverishment of divorced women and their children. Nationally, 25 percent of divorced fathers make no child support payments at all and another 25 percent pay only partially. Divorced men are more likely to meet their car payments than their child support obligations—even though, as one study in the early 1980s found, for two-thirds of them, the amount owed their children is less than their monthly auto loan bill.

When it comes to the collective debt of deadbeat dads, big numbers are routinely thrown around and interpreted in various ways by different interest groups. According to the Office of Child Support Enforcement, 54 percent, $7.6 billion, of the child support due in 1994 was collected, but only 7 percent of that debt had been owed for more than a year. Although enforcement is getting more vigorous, it appears that the longer the clock ticks, the smaller the chance of actually collecting on past debt. These numbers represent only revenues channeled through the federal government; it's hard to get accurate figures because women don't always report child support, paid or unpaid. Collecting child support is more complicated than it seems. It's legal to attach a paycheck to deduct child support payments automatically, but this satisfying remedy is easily dodged when noncustodial parents become self-employed or move out of state. Maine has begun revoking the drivers' licenses of parents who can afford support but don't pay, an innovative strategy that is being adopted in other states.

An added wrinkle is that when dads go deadbeat, ex-wives are often reluctant to press the matter. Some mothers don't want the additional dealings with ex-husbands; some can't afford to go to court; and some have learned that even if a man is ordered to pay, the money may not be forthcoming. When they are pressed for money, ex-husbands often get angry, and to many mothers it isn't worth exposing the children to renewed hostility.

Rosemary's rationale for letting her ex-husband off the hook makes sense. She'd like nothing better than for her ex-husband to fulfill his legal obligation and pay his half of the boys' school tuitions, "but since he doesn't have an income, it's kind of pointless for me to try to enforce it," she explains. Ever the pragmatist, she adds, "My understanding is that if I put him in jail, then my tax dollars will support him. If I leave him out of jail, his girlfriend can feed him. It's not worth going after him, and it's not worth having my kids understand that I'm the one responsible for putting their father in jail, because they probably wouldn't understand that *he's* the one responsible."

Having his driver's license revoked might have worked in the case of Phoebe's ex-husband Howard; certainly no other remedies had any effect. He was a tough nut, contesting the divorce right up to the twenty-three-month state limit and suing for custody of the son he had never fed or changed. Howard had always been a steady worker, but "every time he would start working and I would attach his paycheck, he would quit his job to avoid paying child support," Phoebe explains wearily. Finally she gave up. One Christmas she got laid off, and explaining to her son that she wouldn't be able to get him everything he wanted was devastating for Phoebe. "My son said to me—I choke up when I talk about it—'Mommy, it's okay. It's good that I don't get everything I ask for, 'cause it makes things special when I get them, and it makes me value things more.'" Phoebe now earns enough as a women's clinic administrator to have put such deprivations behind her, but the memory still stings. As she learned, the law can't force a sense of personal responsibility upon an individual.

THE SECOND-FAMILY ISSUE

One reason men default is subsequent financial obligations—or previous ones. An all-too-common consequence of divorce is that Dad bleeps off the screen, especially if he remarries and has more kids. Perhaps Phoebe should have been forewarned by the fact that Howard paid no support to his first wife and three children, who were consequently on welfare. (Phoebe found out about three *other* children, with three different mothers, when Howard asked her to get something out of his wallet and she found a picture of a child, signed "to Daddy, love Tiantha.")

Whether it is more acceptable for a father to default on child support because he is financially strapped or because he has started a second, hungry family is a thorny question. Three out of four divorced people remarry, and data collected by the National Center for Health Statistics show that 88 percent of remarried men had or expected new biological children, or stepchildren, or both. Tory actually benefited from her ex-husband Mickey's remarriage, because his new wife had a stronger sense of his responsibilities toward his son than he did. The only time Tory received support was during Mickey's marriage to this woman, who had children from a previous marriage and was herself receiving it. "She felt it was what should be done; she was a pretty terrific gal," recalls Tory. "I think it was $150 a month, great at the time." The new wife was supportive in other ways as well: during that marriage Tory's son spent summers with his father. When Mickey got divorced again, the support and the visits stopped cold.

Unfortunately the laws need to change for the element of luck to play less of a role in collecting child support. "A truly feminist, pro-child divorce reform would look something like this: dock alimony and child support automatically from the sole or primary wage-earner's paycheck, and let whoever has primary custody of the children keep the house," writes Hanna Rosin, of the *New Republic*. "To punish trophy hunters [men who dump first wives for younger women], force them to support the first set of children at the same living standard as the second." State laws have advanced, with Wisconsin even ensuring mothers tax-supported child support, but most states have a long way to go toward achieving full compliance from delinquent parents. Rosin notes a distasteful reason suspected by many researchers: many legislators are divorced men.

Sadly, in part because parents who don't pay support are more likely to drift out of their children's lives, divorced or never-married fathers who remain equal players in their children's lives are a minority. Forty percent of American children whose parents are divorced have not seen their fathers in the last year. The majority of divorced dads leave the parenting to their ex-wives, which is everyone's loss.

Stepparenting and Blended-Family Issues

The ever-chipper Brady Bunch notwithstanding, it's not easy to combine two families. Nothing is more loaded than issues involving children. Virtually all the arguments Bob, my significant other, and I have revolve around our very different parenting styles—he thinks I'm absurdly doctrinaire, while I find him disastrously permissive, and it's the main reason we don't live together. On the other hand it's clear that we couldn't possibly survive as a couple if we weren't equally weighted by the ballast of children and equally aware of the need to put our respective offspring first if choices must be made. Because we haven't forced togetherness on the kids (though certain group activities are mandatory), both sets accept each other's presence in their lives with pretty good grace. Any one of our four children could easily throw a major wrench into the works, and I appreciate their forbearance. I'm also lucky that Bob loves my kids and sees them as a enriching his life.

Many adults are less generous about the fact that getting involved a second time around is usually a package deal. They may resent the loss of privacy, free time, and disposable income that parenthood entails. The remarriage of a former spouse often provokes a crisis, bring up all sorts of buried feelings and upsetting a delicate balance between families. A decade of research shows that remarriage may or may not benefit children emotionally, though it usually improves their financial situation. In fact the appearance of a stepparent is often another stressful transition for children.

But such new alliances have a great deal to recommend them, especially for the many divorced people whose greatest loss was the pleasure of shared family life, and if the home life that emerges is happy and calm. Remarriage provides stability and boosts the family economy. Children are glad to see Mom or Dad happy again, and relieved that responsibility for the parent's emotional welfare is no longer theirs alone. The fantasy that Mom and Dad will reunite is laid to rest (at least in theory), and the future becomes known and secure.

The Benefits of Stepparenting

A loving, nurturing stepparent is a gift to children, especially when that stepparent fills a hole left by the biological parent. Sophia's ex-husband took custody of his firstborn, a boy, but expressed no further interest in his two daughters. Nor did he pay any support, which is why Sophia got a job, and that's how she met her second husband, though she has since gone on to open her own travel agency. Sophia recognized the crucial importance of introducing her boyfriend and children to each other in a gradual way. "I didn't believe in bringing a man home to my daughters on Thursday and another one home on Sunday," Sophia declares. She did tell Bill about her two daughters right away, but made no move until "he finally said, 'Am I ever going to see these little people? I don't really believe you have them,'" Sophia recalls with a giggle.

"I said, 'Well, this is how I feel: I know it's early, but I need a commitment from you that there is possibly a future somewhere along the line.' He said, 'No problem, this is it,'" and soon after they were married.

Although five years younger than Sophia and childless, Bill decided that he didn't want children of his own, insisting that "two is all we can afford. This way we can give them what they need." He adopted the girls a year after he and their mother got married. "I thank God every day," says Sophia. "He's their father, never stepfather." Working to Sophia's advantage was the fact that her children were only three and five at the time. Older children are slower to bond emotionally with a stepfather and more likely to resent his presence.

Sometimes a stepparent can be the best thing that ever happened to a kid. "She was great," says Celeste, the French teacher, about the stepmother she herself acquired when she was eight. "She never tried to establish herself as my mother, which would have been completely fatal, but she was much nicer to be around than my father, so actually I was grateful to have her there." Celeste's mother remarried almost immediately, and Celeste's stepfather, "for all his faults, of which there are many, was very effusive and doting. I ended up getting a much nicer father in the bargain." Pragmatically describing her parents' divorce as trading one set of problems for another, Celeste admits that it's hard to separate her own beliefs from family mythology, but feels the divorce left her much better off, especially because she was only four at the time. "They hated each other's guts," she says matter-of-factly about her biological parents, "there was no not knowing it. That's not a good atmosphere for a child." Aware of the damage a conflict-ridden home life can inflict, another woman bemoans the fact that her ex-husband's subsequent marriage is troubled too. "My children spend a lot of time there," she explains generously. "I want it to be a stable, happy environment."

Stepparenting can be as rewarding to the adult as it is to the child. Winifred absolutely loved it. "I think I was a great stepmother, and I think they would agree," she says proudly. Marty had three teenage daughters and Winifred "took a lot of cues from him about staying relaxed, giving them the sense that you trust them." After the divorce, Winifred drastically severed every connection with Marty and his family, a painful and somewhat self-destructive choice, "but sometimes you have to have a breakdown to have a breakthrough, you know?" she says ruefully. Two years later she got back in touch with the girls and still sends presents on Christmas and birthdays, although they have never quite gotten over the feeling of having been abandoned and keep their emotional distance. Winifred believes their feelings are certainly justified, but "it's still painful," she says, weeping over the loss. "I think being a stepmother is the most alive I've ever been, because I was being really loving. You give so much of yourself to children. I think I was the best I could be." Certainly the girls benefited from Winifred's love, though as adults they had to redefine the relationship on their own terms.

The Downside of Stepparenting

Children seldom have much influence over a parent's choice of mates, and some fare badly. Their mothers may remarry in haste, for financial reasons or out of loneliness, and may not seriously reflect in advance on their new men's qualifications for fatherhood. Desperate to make the new marriage work, hungry for their own chance at happiness, women may put the requirements of the new relationship ahead of those of the children. Even those who struggle to be fair may feel torn in two by the conflicts of interest that are bound to arise.

There is considerable evidence that stepfathers in particular are often detrimental to family life, most notably as a source of sexual abuse. The incest rate among girls who live with stepfathers is higher than that of those who live with their natural fathers, because the taboo is less strong when there is no biological tie. Also the age gap between stepfather and daughter may be less, and a stepfather is likely to be less invested in a protective role toward her. Theresa's father died when she was nine, and a few years later her mom married a man Theresa wasn't crazy about to begin with. Then he made a sexual advance. "I did tell my mother right away, and he denied it, but after that I never gave him the opportunity," Theresa explains. She was always fully dressed, she never walked up the stairs in front of him, and she wouldn't stay alone in the house with him. "It was a very strange childhood for that,"

Theresa acknowledges, and as an adult she had to work through the consequences in therapy.

Even when nothing as ugly as abuse enters the picture, many people are too impatient or self-centered to be cut out for parenthood, or simply aren't interested in making the requisite sacrifices. Though it needn't be so, for many people stepparenting involves all the thankless tasks of parenting without the attendant satisfactions and fulfillment. Helen is one of those for whom stepparenting is a dead loss. "As a stepmother I feel like I have no authority," she says, "no point of reference where my ideas or my feelings count. I'm out of the loop, neither fish nor fowl." While Helen likes the three kids well enough, she says bluntly, "I hate being a stepmother. I love their father and this is a part of it, but there is no joy in it, and there's no reward." Consequently she's not sorry that the kids live a thousand miles away and visit only on the occasional holiday.

Some of Helen's disaffection for the job can be attributed to the fact that she has no children of her own, but even for experienced parents the subtle, almost contradictory requirements of the job may be difficult to master. When Bob and I went to a family counselor to work out some blended-family issues, I asked her how I could help his daughter to speak up when something is bothering her. "Your role is to back off the edge of the earth," the counselor promptly informed me. "Not *to* the edge of the earth, *off*."

The Ground Rules for Being an Effective Stepparent

In the wake of divorce and remarriage, children are thrust into situations over which they have little control. They struggle with a welter of strong, often conflicting emotions that even adults find hard to handle: guilt, resentment, confusion, anger, fear, relief, helplessness, and more. The transition from one way of life to another is highly stressful. But a few basic rules can ease the passage and lay the foundations for a stable and nurturing family.

1. *Patience, patience, patience.* You and your partner may be in love, but don't expect the children to be equally starry-eyed, least of all on your timetable. It takes time for respect and trust to develop. A stepmother may have taken Mom's place in Dad's bed, but not in the child's heart. A wise stepparent will acknowledge this up front and not try to buy or hurry the child's affections.

2. *Respect the child's concerns and point of view.* A little empathy and understanding work wonders. For example, my eleven-year-old's stepmother recently asked her for permission to take part in the caravan escorting her to the first day at a new school. My daughter welcomed her considerate "stepmonster." (The term is their standing joke, evidence of their mutual ease of mind.)

3. *Agree with your co-parent, in advance, on clear and consistent family rules.* This will probably mean compromising on issues like bedtime, chores, and TV time; choose your battles. Try to retain the rules and routines of the preexisting household that still make sense.

4. *Back each other up in enforcing rules.* Discipline is tricky. It's wise to let the bio-parent be the primary disciplinarian until the children have accepted your presence in their lives. On the other hand, that doesn't mean rolling over and enduring disrespect when children overstep their bounds. The challenge, "You're not my real father" is legitimate, but so is the response, "Even so, in this house my rules need to be obeyed."

5. *Be fair.* All children are injustice collectors, especially when they feel vulnerable, and their interests should get a fair hearing. It's hard to be impartial when children and stepchildren compete or when an angry or unhappy child acts out, but it's crucial. Rosemary's children aren't wild about the fact that their father's girlfriend "has a daughter who's kind of the princess, and my kids are the servants." Though Bob's daughter has the limbs of a high-jumper, he's made it clear that she has to take her turn in the backseat of the mini-van along with my shorter-legged but same-age progeny. I know a teenager who works two jobs to pay school-related expenses his father says he can't help with, and who bitterly resents the color TV in his younger half-brother's room. This constant reminder that son number two has it better rankles whenever this kid spends time at his father's house, and keeps the two boys from being close.

6. *Don't criticize or compete with the child's other parent.* You are partners, not rivals. Competition sets you up against each other, and any smart kid will exploit that situation to his or her own advantage and to the detriment of peace at home. Since you can only lose, don't play. Instead of trying to replace the natural parent or duplicate that role, try to establish your own distinct relation-

ship with the child—and with the other parent. Willa married a man with joint custody of his son and a hostile first wife, who did things like refer her creditors to the new Mrs. Hawkes—Willa. Wisely, Willa was sympathetic, and offered the services of a friend who was a bankruptcy attorney and who helped the first Mrs. Hawkes get her finances in order. This turned the tide, even curing what Willa refers to as "the raggedy-clothes syndrome—you know, 'I'm going to send my kid to their house raggedy so they'll buy clothes'? I bought clothes, and she ran out of raggedy ones. Eventually she realized she wasn't going to get my goat and just stopped, and we ended up friends."

7. *If you can afford it, move.* If not, redecorate. Reminders of the old way of life make new starts more difficult. Turf matters, and neutral turf is a real help.

8. *Do things as a family on a regular basis.* This works best if everyone is involved in the planning. Family meetings are a good place to air problems, work out solutions, and discuss vacations, schedules, and major purchases.

9. *Get support if you need it.* A family counselor can be helpful, especially when a child is troubled or parenting issues are straining the relationship. Stepparenting associations exist in many communities. For the address of the one closest you, call the Stepfamily Association of America in Lincoln, Nebraska, at 800-735-0329. Membership is $35 per family and includes a book, a subscription to a newsletter, and discounts on related resources such as regional conferences.

VIRGINIA
I still am against divorce if at all possible, especially if there's a child. Divorce sucks. But if you gotta do it, then do it. And then celebrate.

Even when everyone in a family is biologically related, even when no divorce comes along to disrupt things, being a decent parent is hard. It's also vastly enriching, in both directions. Children can't have too many adults loving and encouraging them, and whether such support comes from biological family, a "family of choice," teachers, or neighbors doesn't matter in the long run. It shouldn't matter at all.

Divorce affects the life of every child it touches in a different way. At the least it imposes a lesson that is hard to bear at any age: that life is not fair and that virtue matters less than it should. Still, all but a very few of the children in this study have adjusted and are progressing nicely: they're doing well in school, they're maturing normally, and they have good relationships with their peers and other adults. They vigorously disprove the stereotype that kids of divorced parents are inherently disadvantaged, and they have plenty of company.

The mothers in this book know their children are okay, that their own maturity and open-mindedness helped bring the kids through unscathed. Proud and pleased with their children and themselves, they too are ready to go out and play.

Sex
and
Body
Image

*I*s there a woman on the planet who isn't ambivalent, at best, about her body? I've never met Demi Moore or chatted with a supermodel, but I'll bet they too feel the mixture of satisfaction and loathing that characterizes most women's attitudes toward their physical appearance. Pervasive stereotypes—that glasses connote frigidity, or curves mean dumbness, for starters—reinforce women's ambivalence toward their bodies, and are perpetuated by women as energetically as by men. Ever wondered if the "C" in *Cosmopolitan* stood for the cover girl's bra size? Heredity may be destiny, but that's no reason to take it lying down, as the women's magazines gleefully remind us. We'd better hop on that Stairmaster and check out those creepy before-and-after liposuction ads.

Centerfolds notwithstanding, sexuality inhabits the mind more than the body, as the women in this book have learned. Despite our culture's relentless idealization of youth and beauty, the young haven't cornered the market on good sex. There is sex after divorce, and for almost all of these women it is better and more frequent than it had been before or during the marriages described in these pages. That's because the process of divorce involves a psychological passage from repression and self-denial toward acceptance and self-esteem, which is echoed physically and sexually. The journey to self-acceptance can be difficult, though, because women start out so susceptible to outside opinions that they are deficient; ready to believe the worst, they are quick to dismiss a compliment and commit an insult to memory.

Wives and Their Bodies

Marriage results in familiarity, and familiarity does have a way of dampening ardor—after all, nobody needs premarital aids. Ideally, a husband and wife keep the flames going by focusing regularly on their sex life; then growing feelings of security and intimacy deepen their bonds and compensate for any shortcomings in the sexual relationship. This requires commitment, respect, and affection, however, and in their absence, familiarity does indeed breed contempt.

Dealing with a Husband's Sexual Disparagement

As discussed in depth in the first two chapters, a husband's ongoing criticism and belittlement form a dispiriting soundtrack to many marriages. The verbal equivalent of bound feet and whalebone stays, a husband's disparagement maintains his authority. Insults sting, particularly when they refer to women's bodies or sexual attractiveness. Insults stick, too; a single unkind comment is more readily forgiven than forgotten. Willa could not forget the one time her husband Desmond told her she was fat and undesirable, "and that's why he didn't want to have sex with me." Wounded, Willa nevertheless had the good sense to understand that the sexual problems weren't solely of her making, but Desmond consistently dodged any discussion of their sex life. When Willa proposed that they talk about it, "He said, 'I'm not talking about it because you're talking too loud,'" she recalls. "The next time I brought it up I talked real soft, and he still didn't want to talk about it." By cutting his wife off, Desmond maintained control of the conversation; by criticizing her presentation, he reinforced the notion that she, not he, was at fault. Desmond perceived any speech of Willa's as unpleasantly assertive, because what he really wanted from her was silence. By the time he and Willa split up, they hadn't had sex in two years.

Sometimes a husband's distaste manifests itself in gestures, not words, though the message is the same: there's something wrong with your body. Though Gloria describes her husband Lee as "a very good kisser," as soon as they got married, the kissing stopped, and he stopped going down on her too. Mystified, Gloria would ask him what was going on, press him to tell her what had changed, but Lee refused even to talk about their sexual problems. Then one night, after several years of marriage, "He got all worked up and told me that he thought I'd had an infection because I had a bad odor before we got married. I said, 'You married me and [all this time you've been thinking] my crotch stunk?' I said, 'I can't believe that.' I was just dumbfounded," Gloria admits. "God, that was rough." Lee shook his wife's confidence in her essential attractiveness, and by refusing to discuss the problem, much less address it, he made her feel as though she had some sort of incurable disease. Shamed and insulted, Gloria withdrew. The rest of their marriage was characterized by growing sexual tension and intercourse so cold and mechanical Gloria can't even describe it as lovemaking.

Sex During Marriage

A couple's sex life often mirrors the overall health of the relationship. It's a litmus test of sorts, a gauge of how well husband and wife are communicating and getting along overall. At its best, sex is literally "making love," a corporal manifestation of an emotional fusion. At its worst, sex is a weapon aimed at the most vulnerable bits of flesh and psyche. But in any case, whether motivated by love, anger, desire, or self-protection, sexual behavior is highly revealing. Withdrawal is a passive way to manifest anger, and sexual aggression an active manifestation of the same emotion. Looking at sexual patterns in marriages makes it easier to understand why women's sexual attitudes and behaviors changed once the marriages were left behind.

WHEN SEX IS WITHHELD

Orchestrating the needs and desires of two partners over time is one of the toughest balancing acts of married life. Problems in a marriage usually show up in bed, where hidden anger can manifest itself. Regardless of stereotypes of lustful men and gals who just want to be hugged, sometimes the woman wants sex more often than her mate.

Winifred's husband Marty had a lower libido than she did, though she is quick to blame her mishandling of the problem for how infrequently they made love. Winifred draws upon a movie scene to explain her and Marty's sexual relationship; interestingly, she recalls the scene in reverse. "There's a bit in a Woody Allen movie where he's saying [to his therapist], 'We do it all the time, two, three times a week,' and Diane Keaton tells her shrink, 'We never do it enough, only two, three times a week.'" Relating strongly to the Diane Keaton character, Winifred says, "I think Marty and I both knew that's the way we felt." Winifred simply accepted that this was the way it was going to be. For one thing, she was intimidated by her significantly older husband, who was forty to her twenty-six, and she was also dependent on him both financially and emotionally.

Sometimes a partner will hold back out of anger, as Natasha did. A graceful dancer with lovely café-au-lait skin, over the course of her ten-year marriage she became increasingly aware of how her creative partnership with her husband Florian had exploited her. "I cared about him," Natasha says, "but during a lot of the marriage I was very angry, and I would use this withholding of myself to even the score." At the time Natasha's behavior wasn't conscious; "I was just shut down," she says. Sex is now very different because she's no longer passive, which, she says, is "a big change, even from when I had boyfriends before my husband." New self-confidence and a lover who delights in her expressiveness have helped Natasha become far more sexually assertive. Sexual assertiveness, however, is traditionally a male prerogative.

WHEN HE DEMANDS SEX

When it comes to sex, femininity and passivity are almost interchangeable, because sexual hunger is taboo for women. Men, on the other hand, are expected to have urgent sexual needs. Assertiveness is admired, even subtly reinforced with a nod and a wink, because that's the way boys are. Implicit in the marriage contract is the woman's obligation to fulfill her wifely duty, whether or not it's fun for her. "Lie back and think of England," was the inimitable Queen Victoria's sex tip for girls. Ignorant, intimidated, and eager to please, many wives focus completely on satisfying their husbands. Even when those demands are unreasonable, women often submit because they are afraid of offending their husbands, sending them into the arms of another woman, or arousing them to greater violence.

At its most extreme, sexual assertiveness takes the form of rape. Rape is not less brutal when it takes place between husband and wife, but it may seem so to those who believe marriage involves proprietorship, like the husbands of several women in this book, Phoebe, Delores, and Sophia among them. Well into the 1980s it was legal in most states for a husband to force sex upon his wife, and only in late 1993, with North Carolina the last state to concur, was marital rape declared a crime across the United States. It remains a major form or violence against women, but is underreported and seldom prosecuted. Women who are raped keep quiet because they are suspected—and even accuse themselves—of being accomplices to the crime.

Tory, the human resources coordinator, married in 1966, well before the concept of marital rape had entered the public consciousness. A shudder of distaste crosses her face as she recalls her new husband's directives: "He'd say, 'You're my wife. Get undressed. I'm entitled to this.' It was very horrible. And I knew that if I physically resisted, I would be overpowered, even while I was pregnant." For Eileen's ex-husband, sex was a substitute for other forms of intimacy. Instead of talking about his feelings, she recalls, "If he felt bad, he wanted to have sex, if he felt good he wanted to have sex . . ." Her voice trails off. Because Eileen's ex-husband felt that his wife's mood at the time and her needs for cuddling or conversation didn't matter, she recalls feeling "psychologically battered, and just short of physically battered." Eileen chose to remove herself emotionally rather than physically. "I just wasn't even there," she says bitterly. "I'd be kind of staring at the ceiling going, 'I don't understand why we're doing this.'"

Not until many years had passed did it occur to Yolanda, the public relations director with the learning-disabled son, to take issue with her husband's utterly selfish and extremely demanding sexual behavior. As a young bride she didn't believe she had the right to say no, and Steve figured his wife should have sex with him whenever he wanted. Yolanda did her best to accommodate him, but their honeymoon set the unfortunate tone of their sexual relationship. Steve was pretty insatiable sexually, "and I got literally swollen, cracked, bleeding. His solution was to buy all kinds of lubricants, which did not relieve the pain. Here's how naive I was," recalls Yolanda. "I went to see my gynecologist pretty promptly after we got back from this, and all unknowingly entered into the great male conspiracy. I said, 'Here's what happened, what's the solution?' He said, 'Giving him oral sex.'" That solution suited Steve, and for a long time Yolanda went along with it. "It literally took me years to figure out that a good solution would be a whole different approach," she says. Eventually Yolanda began rejecting Steve's advances, after it became clear to her that "he was not trustworthy sexually, in the sense that any opening or relaxing I did would be repaid unkindly, and used against me."

Theresa, the nurse from Naples, Florida, also gave in to her husband's sexual demands for many years, as well as his insistence that she produce a son and heir despite the physical cost to her. David was extremely disappointed when their first child was a girl, and Theresa, who badly wanted more children anyway, ignored a dangerously abnormal Pap smear and a diagnosis of uterine cancer and conceived again almost immediately. The doctor was furious and wanted to take the uterus out while Theresa was on the delivery table. "If this is not a boy, I want that uterus saved," was David's response. When little Camille emerged, "He looked the doctor in the eye and said, 'Save it.'" Although reduced to "that uterus," Theresa had yet to come to terms with how much her body mattered to her husband, and how little he cared about the person who inhabited it.

Surgery six weeks later cured her cancer. Not sure she was still fertile, Theresa unexpectedly got pregnant again. David was very angry when she refused to undergo amniocentesis, because he wanted her to have an abortion if it was a girl. It was. When the baby was one, David slid a piece of paper across the table and said he wanted a written commitment that she would have a baby within two years or he would leave the ten-year marriage. "I'm too old to wait for years and then to search for a new wife to have the son I want," he explained charmingly. "If it's a fourth girl, I'll know at least you tried." Theresa took the piece of paper, wrote on the bottom of it, "I want a divorce," handed it back to him, and said, "'I *want* to get out of this marriage. If you want to find another wife, you better hurry up, you're pushing fifty.'" Pushed to the wall, Theresa finally was able to see herself as something other than the baby maker on which David's approval—and her own sense of self-worth—had rested.

BIRTH-CONTROL ISSUES

The equation of a woman's value with her ability to bear children is a familiar one worldwide. It's a central tenet of the Christian Coalition, which goes hand in hand with their resistance to legal abortion and to sex education. At issue is a woman's sovereignty over her own body. If a woman has access to information and contraception, then she can decide whether and when to have babies. If her value to society is based on a process over which she has control, the balance of power shifts. She becomes less dependent on men, a possibility acknowledged in a prayer offered by a woman at a 1992 Operation Rescue rally: "Oh, please, Lord, break the curse on women's hearts that says we don't need our men. Break that independence." The political implications are considerable: the June 1992 Supreme Court ruling that defended *Roe* v. *Wade* and a woman's right to legal abortion read in part, "The ability of women to participate equally in the economic and social life of the Nation has been facilitated by their ability to control their reproductive lives." It's easy to forget that not until the advent of the Pill in the 1960s was safe, affordable birth control widely available (not that the Pill is all that safe).

The politics of birth control have personal implications as well, and how contraception is handled can reveal a great deal about the inner workings of a marriage. Rosemary and Gil's sex life, quite

energetic during the first five or six years of marriage, took a sub-
stantial turn for the worse when Rosemary, the air traffic controller,
retired her diaphragm. "The thing was, I wanted another child and
he didn't," explains Rosemary, "Although I would never have
sprung it on him, because I think that children need to be wanted
by all their parents, I really didn't want to spend a lot of energy and
money and time preventing something that I didn't really want to
prevent." She was unwilling to subject her body to the health risks
of the Pill or an IUD, and took unilateral control by handing over
to Gil the responsibility to use condoms or get a vasectomy. He was
resentful and angry, and their sex life never recovered.

Whether and when to have children is arguably the most important decision a husband and wife can make, and when they dis-
agree any solution is fraught with tension. Anneke, the romance
writer from Tucson, and her husband George had agreed to start a
family once their finances were in shape. As soon as there was
money in the bank, Anneke several times brought up the subject of
having a baby. Each time George would point out, "We have such a
good life, we can travel any time we want, kids would just tie us
down . . ." Even though she had wanted children very badly,
Anneke finally went along with her husband's program, a decision
she gamely justifies with the rationale that "any children we would
have had would probably have needed therapy from the time they
were five." But later she mentions that both her cousin and her best
friend, who had never even met, made the same comment when she
told them she was planning on getting a divorce. "I said something
about not wanting to hurt him, and they both said, 'Don't worry
about that, because I remember what your eyes looked like when
you said you weren't going to have children.'" At pains to describe
the decision as a joint one, Anneke exemplifies the tendency of
married women to take responsibility on themselves, appropriately
or otherwise. It's one way of making the best of a bad bargain.

Wives Take the Blame for Poor Sex Lives

The habit of assuming not just responsibility but blame comes
easily to women, especially when it relates to sexual problems.
Naiveté and intimidation may cause women to be deferential or
indirect in expressing their needs, behavior husbands often perceive

as deceptive or manipulative. But women want to please their husbands, and may find it especially difficult to communicate about sex, where comments or suggestions are so likely to be interpreted as criticism. Because so much of the male ego is wrapped up in sexual performance, wives may also be reluctant to turn the tables when husbands accuse them of sexual shortcomings.

Anneke, who was loath to press her husband about her desire to have a child, also was not honest with him about her sexual needs. She describes her sex life with George during her twenty-one-year marriage as "sometimes very good, but I faked a lot of orgasms." The faking was "another thing that I did out of love for him, theoretically," she says wryly. George got all excited about an article he'd read describing how women could have multiple orgasms. To make him happy, Anneke faked one. "He was just so delighted and thought it was so wonderful that he could do this," she says, pausing, "and then I was trapped." Indulging her husband right down the line, Anneke was willing to forgo not only motherhood but her own sexual satisfaction.

When a husband finds sex unsatisfying or infrequent, it's easier for both spouses to blame the wife's "frigidity" than to confront the possibility that the husband's lovemaking skills may leave something to be desired. If a woman comes to a marriage sexually inexperienced, she has a particularly hard time rebutting her husband's charge that she is frigid. It is reinforced every time sex is unrewarding for her, and the idea that sexual response is learned—and can be taught—is not one a self-centered lover is likely to pass on. When Susannah, the film archivist, divorced, she admits, "I hadn't felt anything sexual in so long." This was another arena in which Stan, her much more worldly husband, lorded it over her; she had slept with only one other man before marrying at twenty-four. He blamed her "frigidity" for their failure to achieve simultaneous orgasms, "and I can't tell you how incredibly naive I was," Susannah recounts. "It was a terrible thing. I went to doctors and the whole thing to treat this elusive problem that actually required some effort on his part, which he was not willing to make." Stan's abdication of any responsibility for his wife's sexual pleasure was another way of abandoning her. An affair right after the marriage ended did wonders for Susannah; afterward she felt *much* better about herself sexually. "I realized, 'This is some kind of fun.' I

hadn't really experienced the sexual revolution"—this was 1970—
"so I seriously made up for lost time."

Phoebe, the women's clinic administrator, also realized that
"nothing was wrong with me." During her marriage, however, her
husband insistently delivered just the opposite message. A virgin
until she was twenty-one, Phoebe too had only had one lover
before marrying at age twenty-three, and she believed Howard. She
left her marriage feeling "inadequate because my husband would
always degrade me, sexually as well. With him it was 'Wham, bam,
thank you ma'am.'" Nor was sex frequent—"I always teased him,
said my getting pregnant was the Immaculate Conception"—but,
cruelly, Howard made little effort to conceal his numerous affairs.
"Now I laugh, looking back, because it was *him*," says Phoebe
happily. Her post-divorce sex life is very different indeed. "Oh, I
love it. It's wonderful," she says, attributing it to "feeling freer, and
more comfortable in who I was, because I realized all the inhibi-
tions were a result of the affairs, and him telling me that sexually I
wasn't satisfying him."

Life Changes, Body Changes

Just as divorce alters the way women feel about their aptitudes and
abilities, as we have seen throughout this book, it also affects how
they feel about their physical selves. Sometimes it's purely a question of attitude; walking into a room with confidence turns heads,
regardless of how good-looking someone happens to be. Often,
though, the process of cutting loose is accompanied by significant
physical changes. Their scope and nature depend on a number of
factors, including a woman's overall constitution, how stressful the
decision-making and divorce process has been, how she feels about
her body going into the transition, and her age and circumstances.

The women I surveyed were anywhere from nineteen to fifty-
eight when they divorced, with a considerable majority in their
mid- to late thirties at the time. They run the gamut physically,
from svelte, sophisticated types to broad-in-the-beam mothers of
four in easy-fit polyester. For a small but distinct minority, especially for those who were childless, age was definitely a factor in
the timing of their divorces. "I think I wanted to get out of it while

I was still young enough to meet somebody else and have children," says Celeste. Realizing, correctly, that it wasn't going to happen overnight, she figured that it was better to make the move at thirty than at thirty-five. Now thirty-six, she has just gotten engaged.

Laurie and Keiko felt that because they were young and still had a great deal of life ahead of them, they might as well end their marriages sooner rather than later. Thirty when she divorced, Laurie says, "I still felt pretty invincible, that life still had a lot of options for me and I wasn't ready to just settle for this life of quiet desperation." Keiko divorced at thirty-two "because I thought if I waited any longer I wouldn't have the courage, the same amount of confidence." Slender and chic, her story peppered with references to "looking good," the elegant Japanese-American takes justifiable pride in her appearance; her nails are perfectly manicured, her makeup beautifully applied, her shoulder-length hair carefully styled. "I thought it would be harder for me as I pushed forty," she says candidly—harder to end the marriage because it would be harder to attract a man. Her preoccupation with her physical appeal is, however, not typical of the majority of women interviewed. Deciding to end the marriage forced each to confront the distinct possibility of a solitary future, and each chose that possibility over her marital circumstances. The great majority would agree with Nancy, the tableware designer, who said, "My age had nothing to do with it. It was, 'Let's get on with it.'"

Getting in Shape

For many women, beginning to exercise and adopting healthier lifestyles helped them reclaim their bodies much as therapy had done for their psyches. Taking care of one's body is a part of taking responsibility for oneself, and it's an important component in building self-esteem, or what Lorraine refers to in general terms as "all this self-improvement stuff that was going on in my head." For her and other women in this study, looking and feeling better made it easier to envision a different future. Like the subconscious "nesting instinct" that comes over women just before they give birth, a change in outward behavior is often the first sign of an inner change that is just beginning to make itself known. Lorraine began the process of reclaiming her body well before divorce was a conscious option. She began taking long, early-morning walks; at first it was for solitude—"I wanted my own sanity, space, and time, no Lorraine, no Mom, no Mrs. Watson"—but then she started losing weight and feeling good, and kept at it.

Theresa, the nurse and mother of three, went for a complete makeover, though it took more than a haircut to repair the damage systematically inflicted by her husband's regular "report cards," in which her sexual and personal shortcomings were set forth every six months. Though he had gained weight and insisted on wearing a toupee that Theresa found a complete turnoff, David told his wife she was fat and ugly and didn't even dress well. One day a sobbing Theresa called up her sister-in-law Angela, who said, as gently as she could, "Theresa, you *do* look terrible. Your clothes, your hair, your face look terrible. You're sad, sad on the inside and the outside. Can you get back to that young, vibrant, sexy woman you used to be? You're in there, why isn't it out? That's something you have to look inside and find out."

The change needed to come from within, as Theresa's sister-in-law acknowledged, but it began with a few small, tangible steps. When Theresa protested that David barely gave her enough money to buy bread, let alone a new wardrobe, Angela proposed, "Don't buy the bread. You go out and you do your hair, and you do your nails, and you do your face and you buy your clothes, and then when he comes home and there's no food, say, 'There's no food. I had no money.'" Theresa did just that. Then she went for a checkup, started to exercise, began losing the forty pounds gained during three pregnancies in four years, and cut her long hair. "David said, 'I hate it.' I said, 'When my hair was long, did you like it?' He says, 'I hate curly hair.' I said, 'You hate it any way.' He said, 'That's true.' My appearance was *never* satisfactory." Once Theresa realized she should stop trying to conform to David's unattainable ideal, she began finding a look of her own. Once she started looking better, she started feeling more confident and optimistic. Six months into the rehabilitation, when her third child was a year old, she told David she wanted a divorce. Three years after her divorce, her hair a close-cropped cap of dark ringlets, Theresa looks terrific, in shape and in style.

Getting in shape often means a serious change of lifestyle, but as Jodie, the software analyst, discovered, "Things like working out are really important. I hate it, I hate the time it takes, I hate it that getting up the first thing I think of is 'Uggh,' and yet it changes the whole day." She started doing yoga, spends a week each spring and fall at a diet clinic, and has continued to exercise regularly. Extremely health-conscious because she suffers from an autoimmune disorder, Jodie appears in a book about healing yourself called *Spontaneous Healing*, but as the exception: a clear case of someone who would have died without medication. She draws a marital analogy: that without external intervention—in her case, divorce—she could never have gotten well.

Exercising was important for many women, making them feel good, making them look good, and providing a healthy outlet for the stress and anxiety of the divorce process. A woman who is fit and strong feels better equipped to cope with crises and changes. As Laurie says, working out "four or five times a week in aerobics class, jazz class, that was like my therapy." Eileen played a lot of tennis, and went from a size twelve to a size eight. Many divorcing women lose weight, but exercising is not the only reason. Weight loss can also result from illness, poor eating habits, or anxiety. A stranger to exercise, I lost all my post-pregnancy poundage by the less healthful expedient of simply not eating, and when people commented, "You look great," I'd screech, "It's the stress!" I didn't hit the bottle, but some people do take up alcohol or cigarettes during such periods of acute stress. Several women I interviewed mentioned that during the worst times they drank more than usual. "Scotch and cigarettes worked for me," says Celeste, deadpan. "And actually the first meal I made for myself was a big steak." The French teacher also smoked, "because I'd always been sneaking cigarettes throughout my whole marriage. They weren't allowed, so smoking was really fun." Cigarettes had a symbolic significance for Celeste. But she is, after all, a women of the health-conscious nineties; at the same time she started working out, got in much better shape, and quit smoking after meeting the man to whom she is now engaged.

What Celeste was enjoying was the right to treat her body any way she pleased, with no interference from conscience or spouse. Though not physically healthy, such behavior is psychologically so—as long as it is only a stage—because it is part of taking responsibility for oneself. Just as cigarettes represented freedom to

Celeste, alcohol represented dependence to Yolanda. "My husband quite liked my drinking problem," she observes. "He made every drink I ever drank, until I began to feel that I would have to leave him and that my drinking might be used against me. So I stopped, on a dime." Her husband Steve found himself "in a real pickle," in Yolanda's words, because he had to support her sobriety, but "he needed his martinis." AA helped Yolanda, and although she still smokes, she never started drinking again.

Getting Healthy

Getting in shape is often part of getting well, a larger process that operates on both conscious and subconscious levels. Quite a few women in this book suffered from a variety of ailments during their marriages and through the divorce process, while others were literally cured by divorce. As a serious skeptic of the mind-body connection, I was no easy convert to the belief that emotional stress and unhappiness often manifest themselves through physical ailments. But the finding is not mine alone. The California Children of Divorce project found that for women in particular, "many of their somatic symptoms and psychological dysfunctions disappeared during the post-divorce years."

Women often find that when they make that final decision to get a divorce, ailments that may have been plaguing them for years often diminish or even disappear overnight. Women in this study suffered from a variety of maladies during the difficult years before making their move and during the stressful period of separation. Afflictions included panic attacks, insomnia, allergies, migraines, and gynecological problems, and many were symbolically significant.

For example, Megan suffered from gynecological problems during the entire course of her marriage, which happened, not coincidentally, to be a sexually unsatisfying one. After she developed a huge ovarian cyst that required major surgery, and then a breast tumor, her doctor sat her down and said, "*What* is going on with you? You're only twenty-eight, you should not be having all of these physical problems." Megan's job with a growing cable-TV company in Boston was demanding but rewarding, and she's sure that the source of her physical problems was psychological. "I was making myself ill because I was in such turmoil inside," she explains. Since the divorce, she's been fine.

Megan was not the only one to suffer from "female troubles," and to be cured by the divorce itself. "When I was married I started having pain when I was urinating," says Winifred. When a urologist recommended surgery, she responded, "Excuse me? I'm outta here." This was a wise decision, as it turned out; Winifred's symptoms disappeared after she separated.

In Willa's case the stress of ending her marriage was compounded by the fact that she was ending an affair of over twenty years' duration at the same time. One day at the height of this difficult period of separation, she was nearly overcome by a rapid heartbeat and the sensation that her left arm was numb. Terrified, she called her doctor to say she was having a heart attack. "Don't flatter yourself," was his response, but he nevertheless told her he was sending an ambulance. "I said, 'Save your breath, I'm on Third and La Cienega heading for the emergency room.'" Willa landed in the hospital for the first time in her life, but was both surprised and relieved when her doctor labeled it a classic, textbook-case panic attack. Once she acknowledged how much stress she was under and realized that it was manifesting itself physically, she became less anxious, began dealing with the root causes of the stress, and had no more such attacks.

Young and energetic, Susannah didn't begin to have health problems—low blood pressure brought on by exhaustion—until Eliza was born, two years into the marriage. Susannah went back to work at New York's Museum of Modern Art four days later. "It wasn't 'Superwoman, hear me roar,'" she explains almost apologetically. "I had to. I was in the middle of a project, and we needed the money." Nine months later she consulted a neurologist because she had started fainting. (Her blood pressure was 80 over 40.) She kept saying, "I don't think I'm sick, I think I'm just really, really tired." Susannah's physical and mental exhaustion was brought on by the fact that she was entirely responsible for breadwinning, child care, and running the household, and received no emotional sustenance in return. Then her mother took her away for four days, during which she literally did nothing but sleep, "and when I got home, it was as if a veil had lifted. I said, 'What actually am I doing here? What's going on?'"

It took Susannah a few months to figure out the logistics of leaving her husband, during which time the prospect of being alone with him literally made her sick. As soon as she closed her daughter's bedroom door at night, she'd break out in hives, enormous welts, "every kind of rash and itch that you could possibly have," she swears. She would spend the evening in the tub trying not to scratch, and in the morning when Eliza woke up, she'd be fine. At one point Susannah went to a dermatologist. "I said, 'You can't see anything now, but this is what happens,'" she recalls, "and he said, 'Oh, you're having marital problems.'" She hasn't itched since.

It turned out that the cure for Wendy's migraine headaches also happened to be their source. She had suffered from the headaches for years, "maybe a half a dozen a year, and they were totally crippling," she says. Now a cookbook writer, at the time she was a doctor's wife. "There were times when he came home in the middle of the day to give me a shot of Demerol or whatever the magic injection was. I remember thinking, 'What will I do if I'm not married to him?'" She gives a little yelp of delight. "It was the most incredible thing: after the divorce I never had another migraine headache, I never had another one!" An acquaintance of mine had a similar experience. She suffered from chronic insomnia virtually every night for several years before her marriage ended, and systematically tried every remedy on the books, from counting sheep to melatonin. The night she moved out, she slept like a baby, and still does—except on nights that follow upsetting conversations with her ex.

During the divorce itself Francine suffered from colitis, insomnia, and loss of appetite for a few months, but depression was even more debilitating. "There's nothing that takes it away," she says. "I just endured it. I cried. I walked. I called friends. I played music, like a narcotic." When she moved into her own apartment, the depression lifted and her physical health swiftly returned to normal. Delores's health problems, on the other hand, followed her divorce and were exceptionally debilitating, possibly because she had to adjust to a particularly difficult set of circumstances. Her ex-husband had taken the house and car, she had no money, and her family had turned their backs on her for violating the conventions of Cuban culture. Delores recalls, "I was very sick—diabetes, ulcer, I won't bore you—until I realized that the only thing I have is my son. I stopped being sick." Call it mind over matter, call it holistic healing, call it whatever; she, and her companions in these pages, became well. Healthier and happier, these women felt attractive and ready to enter the social and sexual arena.

Women Reclaim Their Bodies

There is a point at which every divorcing woman repossesses her body from her marriage bed and all it symbolizes, and claims it for her own. The transition is full of contradictions familiar to anyone who has missed a warm body on a windy night but also luxuriated in hogging the comforter and sleeping spread-eagled. Olivia wasn't quite ready to give up the physical closeness: after telling her husband she wanted a divorce, she spent the summer in the same house and bed. Still, she exulted in the move to a house of her own that fall.

The period that follows divorce is often one of experimentation. Celibacy may be the right move for wives who were sexually brutalized, while those who were frustrated or repressed may go through a period of sexual experimentation or promiscuity. Others may veer between these two extremes. For many people such exploration is an essential part of the recovery process. There's nothing like a passionate affair to make a woman feel really alive, nothing like the conviction that she's having better sex than anyone she knows to remind her that she is a sensuous and desirable woman.

Transitional Lovers

Quite a few of the women I interviewed made the move from the marriage bed to a bed of their own in the company of a lover. These affairs served an important purpose, providing vital companionship and reassurance during the transition to an unmarried state. Many affairs grew out of friendships that had been platonic during the marriage, and most women kept in touch after the physical relationship had ended. Interestingly, however, *not one* of the women says she left her marriage for the other man. Instead they felt the affair served as a catalyst for what would have come to pass sooner or later.

Toward the end of her marriage Tara had an affair with a man she met through her practice as a massage therapist; paradoxically, the connection improved her sex life with her husband. Forty-seven at the time, she felt "that I was coming into my own. I just felt really juicy, real open, and I brought that home," she says with a smile. "I felt pretty potent." Her fantasy was that she and her lover would spice up each other's marriages, but when it precipitated the end of her marriage she ended the affair at the same time. "I realized the other relationship had been a vehicle," she acknowledges, "that I wasn't leaving Hal for him." For a long time afterward she shut her lover out of her life completely, entering what she calls "a kind of hell of isolation" and a long period of celibacy. It was great, she maintains, giving her a chance to learn a lot about masturbation and about what she was and wasn't interested in doing with men. Now fifty-one, Tara says, "Sex is better now. I'm experiencing all the titillation of being a seventeen- or eighteen-year-old again, with very little of the anxiety. I couldn't care less if I have varicose veins or dimples on my ass. I just don't care."

In Hillary's case the transitional lover was a woman, a black biker chick from the wrong side of the tracks in fact, whose background couldn't have been more different from Hillary's working-class childhood in Lancaster, Pennsylvania. Hillary had gotten married in large part because she and her boyfriend had been sleeping together and it seemed like the next logical step. Sex was never particularly pleasant or satisfying, but Hillary was too naive to know that "it didn't need to be that awful." When she was twenty-two this woman, one of her community college art students, seduced her, "and I had the first good sexual thing I ever had in my life," Hillary recalls with a smile. Hillary wasn't in love, but even now finds talking about this partner exhilarating, eyes widening at the memory. "I was awakened to amazing sexual pleasure by another woman, and a very exotic sort of woman who came from a very different place than I did." Hillary thinks she consciously chose a female lover because of issues with her father that had come up during therapy, and her subsequent long-term relationship was also with a woman. "I guess I'm bisexual," she says. "I've never thought it was odd or weird." After a few years, though, she says with a laugh, "I was beginning to get very horny for penises, really." Six years later she remarried, very happily.

A transitional lover can provide needed reassurance that a woman is physically desirable. The affair may end well before the marriage does, serving as a catalyst by giving a woman confidence that if she leaves, other men will find her attractive. Nonetheless, while transitional lovers served a useful purpose for these women, promiscuity was not the norm among the women I interviewed. The hot-pants divorcée hungrily perched on her barstool is really a specter. A number of newly single women did jump into the dating scene, but wearied rapidly of its dubious charms and worried, sensibly, about the health risks. As they already knew or swiftly learned, brief affairs could be educational and liberating but they were no long-term substitute for coping with solitude or for learning to love out of desire rather than need. As responsible adults, they were after quality, not quantity, and those for whom sex is important found it.

New Sexual Selves

"For a woman, getting divorced often heralds a personal sexual revolution," writes Abigail Trafford, observing that a period of sexual experimentation often enables the divorced person to break away from patterns and scars left by the marital relationship. Not all the women in this book are currently involved in serious relationships, but even those who are unattached aren't sitting at home waiting for the phone to ring. Part of it has to do with being older and more self-assured, more comfortable in their own skins. As depression yields to optimism, women's newfound confidence makes them feel more attractive and sexy.

But there's more to a new sexual self than self-esteem. It's a manifestation of a whole different attitude toward society and its conventions. Leaving a marriage, abandoning the fort, is such a deeply transgressive act that other taboos lose their hold on women who have taken that first bold step. They discover for themselves what Dalma Heyn calls the "power of transgression." They reject the tired premise that only "naughty girls" take the sexual initiative, and refuse to stick to the object-of-desire "nice girl" role; either scenario becomes possible, along with the whole spectrum in between. If sex is still a duty, it is a duty to themselves and the relationship, not to the sexual partner. Mutual pleasure is the goal.

This approach has risks. Sexual assertiveness is not considered feminine. Even the *Playboy* centerfold adopts a posture as coy as Little Bo Peep's, and is expected to prefer romance to sex itself. While men may fantasize about a sexually aggressive woman (for the same reason that women fantasize about the swashbuckling quasi-rapist), when she materializes in the flesh most men are put off by her very boldness. But women who have ended their marriages have already run that risk legally and professionally. That frees them to assert themselves sexually: to express their desires, to explore their partners' fancies, to claim satisfaction as their due. The erotic possibilities of submission become available to either partner. Inventiveness, playfulness, and self-expression become newly possible. With a lover confident enough to handle it, the sexual payoff for both partners is terrific.

Some women in these pages experienced what might even be called a sexual metamorphosis. The most radical example is Rosemary's awakening to her homosexuality. Fifteen years into the marriage, having always thought of herself as a straight woman, Rosemary began to acknowledge that she was drawn to women for more than just emotional attentiveness. In 1990 she became emotionally involved with a woman and decided she couldn't live with a man anymore. She had already stopped sharing her husband's bed, using her bad back and her night shifts at the hospital as excuses. The real reason, of course, was that "I didn't want to have sex with him, I didn't want to lie down next to him. Waking up with someone is a very intimate act, and I didn't want to do that." By the time Rosemary was actually divorced, she was involved in a hot affair with her girlfriend. "How did I feel sexually? Alive," she declares, chuckling. "You do it a lot in the first year or so of a relationship, and if you're lucky you keep doing it a lot."

Shared by a remarkable number of women was the feeling of tapping into a vigorous and unexplored sexual center. As Megan puts it, "I felt transported, like some sexual goddess had taken over me." It was quite a change. Megan has a perfect metaphor for the sexual mismatch between herself and her husband. "It's like going shopping," she explains. "Anthony would be done with all the shopping, and have all the presents wrapped and ready to put them in the trunk of the car, and I was still trying to find a parking space." About the man who was her lover for two years after the divorce, she says with a slow smile, "we went shopping together, I wasn't looking for a parking space, we picked out everything, we wrapped it together. . . . It was hours, it was *great*, just an amazing place to be." It wasn't just the sex, or the fact that this man was very different from her ex-husband. Megan herself had changed: she had learned to take charge of her own sexual satisfaction. For the first time she felt uninhibited, able to say, "Yes, I like this; no, I don't like that" without embarrassment or anxiety.

Not until Celeste began the affair that ended the marriage did she realize "how dowdy and unattractive, how not-in-my-body" she had been feeling. "I felt so much sexier I can't tell you, so that was fabulous. I just felt like Supervixen, like *Whoooaa!*, you know?" she says, cracking up. There is nothing remotely lumpy or dumpy about Celeste, the svelte and vivacious French teacher, who now acknowledges that "I'm in great shape, really not that bad looking after all." Bolstered by this renewed self-confidence, since her divorce Celeste has enjoyed a sex life that she happily describes as "not predictable at all, even within relationships."

Amy, the Oakland mother who has lupus, has taken her relationship to her body one step further, not just accepting it but becoming comfortable with its imperfection. Her body may be not quite as attractive as she would like, but she has never felt so good about herself sexually. "I put on fifteen pounds, and I could care less," she says. "I don't feel worried about my body. I feel I have nothing to hide. I feel completely open and completely in touch with myself sexually. Nothing frightens me, nothing's embarrassing, I want to try things I've never thought to try before." Though she is not currently involved with anyone, Amy's post-divorce relationships have involved more experimentation. She's delighted with this development, and attributes it to the fact that "I feel entitled, as a taker and a giver."

For Marella sex after divorce was more enjoyable because there was no sense of obligation or entrapment. She reveled in the carefree awareness that, in her words, "Gee if this doesn't work, he can go, and I can go, and I won't have to go through a breakup . . ." That freer spirit helped "the whole aura of the sexual act," explains Marella with a smile, white teeth bright against her dark skin. "It's kind of exciting, like, 'Ooh, this is different. I like this better.'"

"It started with the brain," explains Lorraine, the second grade teacher, taken by surprise by her sexual metamorphosis. Exploring her sexuality was the last thing on Lorraine's mind after her divorce; she wanted to eat peanut butter and jelly sandwiches alone in bed and relish her solitude. But the gentle courtship of a man who lived several hundred miles away seemed unthreatening, as did the fact that she didn't have to depend on him for anything. When he called, they'd talk on the phone for hours. Lorraine didn't even agree to go to dinner with him until three months had passed, and it wasn't until some time after that that they became lovers. "The sex thing, though, geez!" she exclaims.

> I look at him and my eyes are open—and this is a first, instead of the lights being off, my eyes shut—and this is a whole different thing. And you know what it's based on? Trust. Caring. Faith that the person is not going to malign you and expect you to be functional. I look at him and [see] someone who cares so much about me and respects me for my head and my caring ways, and values my opinion (and I do his, you see, it's not just one way). It makes me free, free to say, "Let's try this. Let's do this." And he says, "What can I do for you, tell me what you would like?" and to *him* this is very new, because he was a once-every-three-months guy, because his was not a happy marriage either.

These sexual transformations cannot be attributed solely to maturity, or to the thrill of novelty after a decade snoring next to Dagwood. In choosing to end their marriages, these women have defied gender expectations and stepped out of the loop of self-denial. They've taken responsibility for their own pleasure. This doesn't necessarily mean being smeared in maple syrup or jumping into orgies, but rather bringing all of themselves into bed—or anywhere else—and looking for more equal, active, fulfilling partnerships. Those are ingredients for great sex.

Some Women Need More Time

Following separation, women in this study ran the gamut from sexually voracious to completely shut down and celibate. The women who chose not to have sex did so for very different reasons.

If recently divorced, some are too busy getting their own lives together to devote energy to sustaining a relationship. Older women may feel sex is a less urgent concern, and don't wish to spend their time nursing an elderly mate. Esther, who at sixty-nine has been living alone in West Palm Beach for fifteen years, says honestly, "I would love to have a man friend. In my bed, I'm not quite sure." Still other women who feel sexually turned off post-divorce may have been sexually ill-treated in their childhoods or their marriages, or both.

Theresa, the nurse, was raised by a mother who headed off her abusive stepfather but whose sole advice on the sexual front was "Don't let men touch you." It's small wonder that after enduring sex on her husband's derogatory and demanding terms for ten years, having two surgeries for uterine cancer, and giving birth to three daughters in four years, she describes herself as "turned off. I didn't want to know about any part of my body from the waist down." Without realizing it, she switches to the present tense: "I don't want anybody even looking at my lower body, I don't want anybody sucking my breasts. I don't want to be touched." Three years have passed since her divorce, and she's beginning to think of herself as a sexual being again. "It's starting to come back," she says shyly. "I feel attractive."

It's important to keep in mind, however, that celibacy doesn't rule out sensuality. At the time of her divorce Winifred had had only two lovers. Then thirty-five, she described herself sexually as feeling "like I was ninety-nine years older." In the twelve years that have passed, Winifred has had some memorable sexual episodes, and also some relationships with "people I really liked but the sex was nice but not incredible, which surprised me," and some lengthy periods of celibacy "because I couldn't find anybody I felt comfortable sleeping with." But that didn't mean giving up all the pleasures of the flesh. "I felt, 'Okay, I'm a sensual person. I'm going to take bubble baths, and I'm going to sleep naked, and not give up feeling that I have a natural sense of enjoying all kinds of sexuality,'" explains Winifred. "It's all sex. Touching somebody is sex." Winifred has worked to keep her basic sense of wonder intact and her expectations high, and feels that she's been more fortunate sexually than most women. She has chosen her circumstances, and she's comfortable with them.

Beyond Body

The tyranny of cellulite and crows' feet is powerful in a culture that
values youth and beauty over intelligence and experience, especially
in women. "An ugly woman has no credibility in our society,"
declares Anne Wilson Schaef. "No one even wants to look at her,
much less listen to what she has to say!" Part of the process of self-
acceptance lies in rejecting that notion of worthlessness and coming
to terms with the body, warts and all. Aging forces the issue. As
flesh creases and sags, keeping things firm grows harder and eventually impossible, as a sign outside a Body Shop in the London air-
port ironically pointed out: "To avoid wrinkles," it read, "NEVER
SMILE AGAIN." Having decided to live by their own standards externally as well as internally, the women in this book are smiling more,
and more widely, than ever.

The standards of female beauty may have been set by men, but it
takes two to enforce them. One's lover's tastes do matter, of course,
and it's fun to make an effort to please a man and see it pay off, but
it is liberating when conformity to someone else's standards is no
longer what matters most. Gloria used to be governed by those high-
maintenance standards. "I used to wear all this makeup, and they
called me Big Hair, and my husband would come in the house and I'd
hurry to put my lipstick on, okay?" she says, slightly defensively.
Seeing herself as independent of a man's approval hasn't come naturally. Gloria admits that she still sees security in a wedding ring, and
that she's got to get over it, "because men do not complete me."

She's already come a long way, and the change is reflected very
literally in her appearance. For women who had "let themselves
go" during the course of unhappy marriages, reclaiming themselves
may mean buying new clothes or investing in a professional
makeover; for others, like Gloria, it may mean just the opposite.
She's tossed the eyeliner and the deodorant and has even shown up
at the store in her gardening clothes with dirty nails, a radical
departure for someone once so appearance-conscious. "I don't
want to play these little, pretty games. I want to just be strong, and
rugged," she explains. "I even went through a phase, after Lee hit
me, when I didn't shave my legs or my armpits. I'm not going to rat
my hair and spray it all day, and be putting my lipstick on for anyone. It's like I am who I am."

For the women in this book meeting a man who finds fault in her appearance no longer results in a self-criticism session in front of the mirror but the realization that the man may not be worth going out with. If he declares that short hair is dykey, he'll never have the pleasure of learning otherwise. In her mind, he, not she, is the one who becomes undesirable. Nor is her liberation limited to freedom from crash diets and migrating implants. Unattached, a woman is sexually accessible in a way that a wife is not; divorced, she appears experienced and self-reliant. This is a powerful combo, rendering her both exciting and threatening. But it's not a problem for a woman who isn't ambivalent about the personal power she *knows* she possesses, or for the man who wants a strong and independent companion.

CHAPTER 8

New Relationships

A friend of mine is an actor in her mid-forties who has never married, though not for lack of suitors. Trying to goad her into a rage for an upcoming scene, a director finally mustered his ultimate insult: "Spinster!" he jeered. "My jaw dropped," recalls Amity, "not at the comment, but because he really thought this was the most wounding thing he could come up with."

Conventional wisdom dictates that no sensible woman would choose not to be married. My own reaction to a comment of my mother's showed me just how firmly entrenched the notion remains. When I mentioned that my boyfriend Bob and I just might tie the knot someday, she said, "*You?* I can't imagine you ever marrying again." Was I pleased that she had seen into my liberated soul? No, I was offended—by the inference that I wasn't "marriage material," that no man would pick me. I swiftly came to my senses, and realized that she had actually meant it as a compliment. But it was sobering and embarrassing to realize that a chunk of my self-image was *still* hanging out in Cinderella-land.

Do Women Need Men More Than Fish Need Bicycles?

A feminist slogan of the 1970s was "A woman without a man is like a fish without a bicycle." Whether that's true or not, "bicycles" have always been a popular way to get around. To love and be loved is a natural human desire, and rejecting the possibility of romance means sealing oneself off from a great adventure. Women derive much of their identity from their ability to make and maintain intimate relationships. These attachments create community, give life meaning, and suffuse it with happiness, and women are right to cherish them.

However, our society values independence and equates reliance on others with weakness or neurosis. This puts women at a tremendous disadvantage. Economically, women *are* far more dependent on men than vice versa largely because of the wage gap, especially if they have young children. Socially and psychologically, women are urged to define themselves through their relationships, yet when they do so they are accused of being needy. But to interpret women's desire for attachment as weakness is to miss the whole point. If women were innately ill-equipped to be self-reliant, widows would be more depressed than widowers; the opposite is the case. If intimacy guaranteed happiness, married women wouldn't be the most depressed segment of the population, far more unhappy than single women and single or married men. Women who have ended their marriages have had to learn to take care of themselves, and this newfound sense of independence makes them confident and happy.

LORRAINE
I'll tell you, I wouldn't be in an involved relationship again if it weren't really good stuff. There's a wonderful quote: "It takes a mighty fine man to be better than none."

Bicycles Are Nifty, But They're Not the Only Way to Get Around

In or out of marriage, a woman has to overcome the belief that she is incomplete without a man; otherwise she won't develop the solid self-image that is the basis of any healthy relationship. Despite the women's movement, many women still hear that nagging voice: "If I don't have a boyfriend, I must be fat and ugly. If I don't have a boyfriend, I have no identity," in the words of writer Leora Tanenbaum, who goes on to drive home the underlying point: "this sentiment will persist as long as women continue to believe that a good marriage is the surest path to a happy and fulfilled life . . . that we are only half a person searching for our better half."

The struggle to achieve a sense of self that isn't based on the opinion of others has been waged by every woman in this book, sometimes joyously and other times under duress. It's been particularly difficult for Sharon, the pension fund manager from Kansas City, who summons up the dramatic image of marriage as a procrustean bed, as she describes it, "a medieval Catholic device of punishment, where they would lie you in this bed and however much of you stuck over the end of the bed would be cut off, toes, knees, whatever. A lot of times we find ourselves in relationships where we are basically willing to gnaw off our feet, cut off our legs, to fit into that procrustean bed, because at least we're in a bed," she observes wryly.

After divorcing at twenty-eight, Sharon was without a man in her life for the first time, and she found it very frightening. "Not having the constant attention of a man kind of chipped away at my own neurotic self-esteem issues," she says, calling it "am-I-damaged-goods kind of thinking, am I lovable, you know?" In fact, she had a transitional lover and has been in a number of subsequent relationships, including a two-year live-in one that just ended. Now thirty-five, Sharon claims "a stronger sense of self," but still relies on the attention of men for much of her self-esteem. She's not unaware of the irony, of the tendency to choose weaker partners because the sense of control makes her feel more secure. "But ultimately I lose respect for the person, because of the very thing that makes me feel comfortable," she admits. Slowly, bravely, Sharon is getting to a place where solitude or an egalitarian relationship are both acceptable scenarios. Whether or not she falls in love again, whether or not she makes it legal, she will never again feel as vulnerable as she did when her marriage ended.

PHOEBE

Don't count on the rest of the world to give you support, to save you. You have to value yourself, to start believing that you are somebody. So many of my friends would say "When are you going to remarry? or "You're never going to remarry." But you know what? My response to them was, "It's okay if I never remarry."

Divorced Women Reconnect Out of Desire, Not Need

When it comes to new relationships, women who have gone through divorce often find that the experience leaves them not the least bit desperate. Most would like to be in a committed relationship, but they don't need to be. Stripped of their happily-ever-after illusions, the women in these pages have rejected the popular equation of femininity with childlike dependency. They're not interested in men who find learned helplessness attractive, and they're not the least bit helpless. Nor are they martyrs, a relief to the men who aren't interested in being partners with a saint. Eileen would like a future companion "to be supportive of me but not necessarily support me. And to care about me but not be responsible for me, you know?"

Tory, the human resources director, was even more explicit about what she wanted from a man when a pivotal ten-year relationship was breaking up. She called her therapist and put it to him plainly.

I said to him, "You know what I don't understand? I don't want someone to father my son. I don't want someone to pay my bills. I don't want someone to take care of me. All I want is someone to be my friend, and my partner." He turned to me and he said, "Men know how to father. They know how to pay bills, they know how to earn a living, they know how to appear in public at your side. Those are the easy things that you're saying have no value. You're asking for the very most difficult thing that men don't know how to do."

Because his needs were being perfectly well met, Tory's significant other wasn't interested in rethinking his role. But Tory aspired to a more complex and fulfilling relationship with the man she loved.

Being alone is no longer scary for women who have experienced it. Some may even embrace their time alone, the physical, emotional, and psychological solitude. As Yolanda, the public relations director with the learning-disabled son, puts it:

> I have to say, in retro mode, that the sad fact is that on the whole, women are grown-ups and men are not. It's tough, because one of the things that is empowering about being single and not feeling desperate about it is that you know you will not form a relationship because you *need* to, you will form it because you want to. That's very different from the way most of us get married the first time. I never want to need somebody as a dependent again, and I never will. It will be a pleasure if there's somebody in my life who can be a support system for me, and I know how good I am at being a support system for other people, but the terms are very different if there's no desperation and no need. It's free choice, adult choice, that's new.

Yolanda smiles. "Unfortunately the person I've been seeing for the last year or so is a little bit of an all-bed-no-breakfast kind of guy, but I feel fine about that because I'm making the rules now. And he's okay with that; if he weren't, he's a grown-up, he gets to go away," she explains. "I'm starting to feel like I'd like to meet somebody who isn't a schlemiel *and* is fun in bed, you know, but no rush."

The women in this book have experienced the damage wrought by intimacy between people who are not equals, and have stepped out of that loop. They are learning to distinguish between healthy interdependence and damaging complicity, though heterosexual models are few and far between. They reject relationships in which they have more at stake than their partners, or which involve leaving their identities behind. Whether they are in a relationship is no longer the cornerstone of their sense of self. They have learned that being alone and loneliness are not the same thing. The prospect of solitude no longer terrorizes them, and many have even come to embrace it.

DELORES

Is this person going to add to what I have today? Or is he going to take from me what I have accomplished? If he's going to take, I don't even waste a minute, I don't care if he's a millionaire or the king, because the freedom I have of doing what I want to do and not being controlled, I love it.

Yes, But Are There Enough Bicycles to Go Around?

Although it was statistically flawed and demographically limited, much was made of a 1986 Harvard-Yale study reporting that most single women over age thirty-five can kiss good-bye their chances of ever getting married. Critic Katha Pollitt points out that while the study has been regarded as an astonishing new development, for over a hundred years there have been more women who are educated and independent than there have been men who wanted to marry them. "Then, as now, blame for her unmarried state was divided between demographics, which made millions of women 'superfluous,' and feminism, which spoiled them for the 'sacrifice' of marriage," writes Pollitt. Contrary to the popular image of a crazed female populace wrestling with the "man shortage," many of that larger number of women who remain unmarried do so by choice, or prefer cohabitation to matrimony.

Statistics show that three out of four divorced Americans do remarry. For women the typical interval between remarriages is four years; for men it's three years, and they marry women who are, on average, four years younger than the ex-wife. The remarriage rate is dropping, though. Anne Wilson Schaef puts a novel spin on the fact that fewer newly divorced women are remarrying promptly, unlike their male counterparts who speedily tie the knot again. "Surprisingly, *women* are becoming the gypsies while *men* are becoming the nesters!" she theorizes, a little breathlessly. "In fact I wonder if men have not tried to convince us that we are natural nesters so that we will provide them with nests."

The fact is that both age and gender do affect remarriage rates for women. There is a 76 percent likelihood that a divorced woman in her twenties will remarry; the percentage drops to 56 percent if she is in her thirties, to 32 percent if she is in her forties, and to 12 percent if she is fifty or older. The rate drops as people age, especially for women; men between the ages of forty and forty-four are twice as likely to remarry as their female counterparts. Another factor rendering remarriage less likely is income: the greater her professional success, the greater the likelihood of divorce and the less likely she is to remarry. These figures reflect the fact that two of the main motivators for marriage—money and reproduction—simply don't apply to older and well-established women, though they are often seized upon as cause for hand-wringing.

"I didn't even consider the dating game," admits Willa, a television producer in her late forties who has run smack into a cultural double standard that is especially acute for successful African-American women like her. The more well-off and capable a man, the larger the pool of eligible women in which he can troll. It's the opposite for a professional woman, because, unlike a man, she is expected to marry "up" the economic ladder. The older, better educated, and more competent she is, the fewer "appropriate" candidates are out there. The obstacles are real; Willa doesn't have a date for the upcoming wedding of a colleague, and she's not happy about it. "I didn't think about the implications of my age," she admits. "I'm suffering from something I never imagined: men like me, but they're intimidated by me!"

Willa and other women in her situation need to relinquish the idea of "marrying up" and consider dating younger and less well-established people, just as their male counterparts do. Snobbery, both social and intellectual, can blind a woman to potential mates, but it's as foolish to assume that all professional men are likely to be professionally supportive as it is to assume that all working-class men are likely to be socially conservative. In fact, younger men tend to be less invested in the status quo and more open to letting go of traditional male-female roles. Both the man and the woman have to put aside their own pre-existing expectations and be willing to defy those of the community. That calls for confidence and open-mindedness, but those qualities are a must in any potential mate.

Suspecting every man who comes along of being a sexist pig doesn't work to a woman's advantage, and neither does dating anything that can walk. Especially after coming through a bad marriage, a woman should be picky—*about the right things*. A friend's mother who was divorced in the 1970s and is now in her late fifties doesn't even consider men in her age group, whom she finds childish and tiresome. They've never had to grow up because of all the women catering to them in the dumb and deferential manner recommended by all those how-to-catch-a-man books. The advice in these manuals boils down to a few simple directives: (1) you have to have a man; (2) don't be so picky; and (3) catch one by playing the old-fashioned, girlish game.

The recent best-selling dating guide called *The Rules* is a perfect example. Don't call him, it advises. Don't initiate conversation, let alone sex. Let Mr. Possible take the lead and run with it and he just might turn into what every woman needs: Mr. Right and a wedding ring. In other words, be passive and manipulative and you'll get what you think you want. No wonder Rule number 31 is "Don't discuss 'The Rules' with your therapist." These directives play out nicely for dominating husbands and subservient wives, especially if the men don't abuse their authority. But women who live like this never grow up, and divorced women have been there and done that.

Life as a grown-up is more complicated and much more interesting. It also calls upon women to stick to their guns and not revise their basic natures to suit the first guy who comes along. Amy, who is a single mother and runs a small family business, had been dating a man with leisurely habits and an independent income. One day he looked at her and said, "'You know, Amy, you do too many things, you're always on the go, you wear too many hats.' He went on and on," recalls Amy. "I turned around and said, 'And you're the hat that has to go.' It just, like, slipped right out." Like the other women in this book, Amy was no longer willing to compromise herself just to have a date.

What It's Like to Be Single Again

The romantic landscape has certainly changed. Sex was never "safe," but now it can be lethal. Often children have complicated the picture and expanded the waistband. Childless women may find their biological clocks ticking louder than the violinist at the next table, or discover that attractive men come with bratty teenagers in tow. Even the vocabulary is unfamiliar. Do people still "date"? What do you call your forty-two-year-old "boyfriend"? What is a dental dam, anyway? "It was strange," admits Laurie. "I hadn't been single in a long time. I felt socially retarded, I didn't know the rules."

What Laurie had to figure out for herself is that the old rules work to women's disadvantage and there aren't any new ones. Their absence is ultimately a good thing, but it makes the transition all the more intimidating and bewildering. Divorcing women of all ages are ambivalent about their newly ambiguous social status. They miss the companionship of marriage, of having someone to come home to, and that loss does not abate with time. It is reinforced at every turn by the astonishingly persistent notion that a fulfilled life is one with a man at its center. It takes not just courage but imagination to envision an alternative.

But as with other fears, women's most profound anxieties were seldom justified in retrospect. Single life turns out to hold many pleasures, largely unanticipated especially by women who have never before made a home for themselves. My high point was picking out a coffee table without having to consult a soul as to style or price, which probably explains why it's one of my favorite possessions. For other women in this book, the perks included hanging whatever they wanted to on the walls, going to bed when they felt like it, and not having to make dinner, or conversation, or nice.

Being Single Is Exhilarating

"People say, 'You're not dating anyone, aren't you lonely?' I say, 'No.'"

Megan, the woman from Boston who works in cable television, behaved very differently with the man she started seeing after her divorce, with whom she was very much in love. "We talked about everything. I was much more able to state my differences, to say, 'Well this is part of what I am,'" she explains. "To be an individual was much easier, because I realized that's what I had to do with my whole life, be comfortable with me and be true to myself." They broke up six months ago, after two and a half years together. "People say, 'You're not dating anyone, aren't you lonely?'" says Megan. "I say, 'No.' I totally feel like a whole person, I'm excited about *my* life." Thirty-three, she loves being single and is confident that someday someone will come into her life with whom she'll feel confident enough to start a family. "I feel that the world is at my fingertips, that I know Megan much better, that I know what I like and what I don't like. I know what I would tolerate and not tolerate in a relationship. And being single I have the freedom to make all the choices for me."

"I have a social life! I got divorced!!"

Because now she can see all the kooky friends her husband never liked in the first place, and because her ex-husband is willing to baby-sit their son Asher, and so are Asher's grandparents, Keiko quips, "I have a social life! I got divorced!!" But the number of houses in her conservative Seattle suburb to which she's no longer invited hasn't escaped her notice. "There are people who will not ask me to dinner parties because I'm a single woman," she comments, and she knows why: "I don't come in acting like the down-trodden divorcée, and that's what these women *want*."

Being Single Is Complicated

"What must be understood with anybody I'm dating is, 'I have a kid, and we're a package.'"

The lesson of *The Good Mother*, both a best-selling book and a movie—that a mother could lose custody if she was careless enough to have a sex life—is a terrifying one. Moms aren't supposed have sex, especially divorced moms, who are already skating on thin moral ice. Furthermore, there's no protocol. "I get very nervous in a way, because I don't know how to do it yet," admits Nancy, the tableware designer in New York, who started dating a couple of months after the papers were signed. "What are the rules? Do you have these guys over in front of your kid? I'm still kind of feeling that one out." She makes it perfectly clear to new friends where her priorities lie, telling them outright, "I'm a single parent, and this kid is *real* important to me. We're a package. I'm not going to stick this kid with baby-sitters all the time, so if you want to be in our lives, get used to this little five-year-old." Nancy reports that "a lot of people, in fact a lot of men, I have found, surprisingly, are very supportive and respectful of it."

Nancy felt so isolated in her marriage that she says she always felt single, so when she got divorced, "I think I just felt relief." The adjustment to single life was further eased by the fact that, in her blunt assessment, "I have been operating as a single parent since the day this child was born." Not long ago Nancy broke up with a man she'd been seeing for some time because he wanted to move in. "I had to fight so hard for my independence that I would be very hard pressed to let it go," she explains. That doesn't mean she's given up on the possibility of romance, only that she's not pinning her plans on Mr. Right's arrival at her doorstep.

> I still think there's going to be this Prince Charming out here, you know, I do. I adore men. But I think the type of men I went for before were probably wrong for me. And there're a lot of good guys out there, but I just didn't see 'em. But I'm not in any great crazy search for someone. I just feel that when I'm ready, hopefully it'll happen. If it doesn't, then fine.

"I want him to be lonely and say, 'Look what I lost.'"

Although a woman may have done the leaving, and although her ex-husband may have been a complete and utter asshole, there's no guarantee that he won't find someone new before she does. It certainly seems men tend to hook up again fast, maybe because the pool of suitable companions is larger and perhaps because they seem to be temperamentally less suited for solitude. Accepting the fact that an ex has paired off first can be very hard for ex-wives.

Three years after the divorce, Theresa, the nurse with three daughters, found herself alone in David's apartment for some reason. "Here he is living in this beautiful apartment, hasn't paid me my maintenance this month because he moved and doesn't have any money," she notes dryly. "There are two wine glasses in the dish rack, champagne in the fridge, he's got this young girl's negligée hanging over the bedpost, he sees his kids just enough to show that he loves them, and he walks away and I've got the hard work." Theresa continues frankly, "I want him to suffer. I want him to be lonely and say, 'Look what I lost. I was a bitch to this woman. I lost her love. I lost her affection. She worked like a dog to make this marriage work, this house work, and I didn't appreciate it.'"

Theresa came away upset and resentful, until she asked herself, "Okay, Theresa, if you're so angry, so jealous, would you like to take this girl's place? Are you sorry?" The answer is always no. She still suffers bouts of depression, "but even if I'm down or lonely, I *know* this was the right decision. As soon as he was gone I could take a deep breath, and that's what brings me back up." Theresa remembers checking for second thoughts as she watched David pack up his things. She asked herself a rhetorical question: "'Let's think: when I'm sixty, no kids are here, do I want anything to do with this person?' The answer was no. When I get to be sixty, this is the last person I even want to have lunch with."

Being Single Means Shelving Starry-Eyed Fantasies

"I still want to get married, in many respects, except that I think I like the idea of it; the reality is something else."

The rosy glow of romance is what gives marriage much of its allure—all of its allure, according to Carolyn Heilbrun, who writes that focusing on romance is necessary "not only to give form to male adventures and female lives but to maintain marriage itself. For if marriage is seen without its romantic aspects, it ceases to be attractive to its female half." Yet the pull remains strong, even for women who are well aware of the contradictions involved. "I *still* want to get married, in many respects, except that I think I like the idea of it; the reality is something else," admits Ginny, the Rochester art director, freckled brow furrowed. In the beginning she didn't like being single at all, but she has grown accustomed to feeling "much more powerful than people that are married," which she attributes to "the opportunity for choice. I didn't feel I had all the choice when I was married that I do now." Ginny figures remarriage may or may not play a part in her future, depending on her stage of life. "In eight or nine years my youngest will be done with high school," she points out. "That'll be another significant opportunity for change, a chance to do something really radical."

For Amy part of deciding to divorce was acknowledging that she was mourning an illusion, not a reality. Coming to terms with being single involved exactly the same process. "When I first got separated and lived alone, I thought, 'I really want to be married again. I really want that thing I never had.' Simple things—my daughter jumping into bed with me and my husband and feeling so happy and it's so wonderful!" acknowledges Amy. But she soon realized that what she was yearning for was a television fantasy, and that she and Martin had only shared "those moments of absolute ecstasy," as she describes them, until Cara was about eight months old, though they stayed together for the baby's sake for four more unhappy years. Two years later Amy feels fine about being single. "I don't have a problem—I don't feel that I should be married or I should be with somebody. I haven't minded being alone." She feels that remarriage is not only not a goal, but something she's moving further away from.

Feelings About Being Single Change All the Time

"I hated being single. I still hate it. Even though I love it."

Winifred was divorced thirteen years ago and would like to remarry. She says, "I hated being single. I still hate it. Even though I love it." She misses her stepdaughters and never had children of her own. Interestingly, if she does pair up again, she'd like to "marry somebody like Marty again, and hope that I would do it better this time. That is a really honest answer." "Like Marty" translates into "an incredibly bright, incredibly verbal and articulate, fun, humorous, Jewish, successful man." With "Marty II," Winifred would be more assertive; with her ex, she was "a scaredy-cat. If he said, 'Jump,' I would say, 'How high?' or apologize." Winifred says what would change this time around would be she way *she* related. "I'd say, 'Don't tell me to jump, Marty. Stop being a jerk. Relax.'"

"I had a horrible marriage, but I liked being married."

Virginia shares Winifred's ambivalence about her marital status. She felt "fantastic, absolutely fantastic," about becoming single, but six years later she's tired of being a single mother and ready for a committed relationship. "I liked being married," says Virginia. "I had a horrible marriage, but I liked the structure of it, the one-on-oneness throughout the day, on a daily basis." She shakes her head ruefully at the apparent contradiction, and makes it clear that her first condition for a new mate would be pragmatic: "that he can deal with the world in some sense. You know," she elaborates with a laugh, "bring in a little *money?*" Her goal would be "to maintain the identity I had as a single parent and integrate that into the married identity." That's a tall order, she knows, but she doesn't think it's an impossible one.

"I started being more comfortable with being single, maybe too comfortable."

It's easy to get accustomed, emotionally and logistically, to not having to accommodate the demands of a partner. Phoebe too got used to being on her own; it allowed her to focus single-mindedly on her child and her professional advancement, and she relished the luxury. That's not how she felt right out of the gate, though. "I was scared," she admits, laughing. "Was I going to meet someone who would not only be a good partner to me but also a good father to my child?"

After the interminable two years of waiting for her divorce to come though were over, Phoebe's girlfriend said, "Hon, it's time. You've got to rejoin the world." Men were interested, but as a single mother who was going to college and working full-time, Phoebe wasn't all that available, logistically or emotionally. "Men were like, 'Come on, let's go out, we'll do dinner.' They'd say, 'You're obviously not interested in a relationship,' and guess what? At that time, I wasn't. It was my time for me. I was doing something that was very important to me, and I sacrificed at all costs to do it." She started being more comfortable with being single, "maybe too comfortable with it," she admits. "That's why I procrastinated so long about getting remarried."

At forty, single for thirteen years and very happy with her job running a busy women's clinic, Phoebe grudgingly agreed to a blind date set up by her father. She kept saying, "No, Dad, I'm not interested," until the day her father said, "Look—here's the phone," and handed her the receiver with the guy on the line. They dated for a year before marrying a year ago. "I'm a different person now, very confident, very comfortable in who I am," she says, "and because I'm different, I approached my second marriage differently. I have a husband who appreciates who I am. He's very supportive of anything and everything that I want to do. It's really wonderful that he recognizes that I have my own dreams and goals to achieve in my lifetime, and he has his." An independent and emotionally stable person, over the years Phoebe has adapted to circumstances both welcome and unwelcome.

Why Women Remarry

Noting that over 90 percent of all Americans do marry, anthropologist Helen Fisher observes, "Marriage is a badge of *Homo sapiens*.

To bond is human." Divorced people seem equally subject to this biological imperative, confirming Samuel Johnson's description of marriage as "the triumph of hope over experience." For some women, remarriage was the quickest route back to financial security, though no woman I interviewed gave that as her main motive. Some wanted to have children. Some had sworn off love, only to be ambushed by it in the end.

Depressingly, about a quarter of all second marriages break up within five years, and the risk of breaking up increases with each remarriage. Some 60 to 85 percent of remarriages—including second, third, and fourth ones—end in divorce within ten years, not that that seems to deter people. The stress caused by combining two families is one factor; another is the regrettable tendency to remarry just another version of spouse number one; and still another is the fact that people who have weathered one divorce are likely to be less tolerant of marital misery a second, or third, time around. A more heartening statistic, however, is that the older the bride and groom, the more likely that the marriage will endure. Thirty percent of the women interviewed for this book had remarried; 4 percent had subsequently divorced a second time; and an additional 6 percent, myself included, were unmarried but in a committed, long-term relationship.

Whatever their individual circumstances, women interviewed for this book had different expectations of marriage the next time around. They were more realistic and far more articulate about what they wanted from their mates and what compromises they were and were not willing to make. As a result, they were by and large much happier than they had been the first time around. Success in remarriage, as Abigail Trafford puts it, "depends on a successful divorce from the old marriage and the successful emergence of self."

Women Remarry to Have a Stable Family Life

"He said right off, 'I understand it's a package of three.'"

Like most divorced people, Laurie remarried within three years of her divorce. While deliberating whether to move in with her boyfriend, Laurie took an adult ed course called "Parenting

and Stepfamilies," which was run by a psychologist she liked a
lot.

> At one point I said, "I don't really believe this guy when he says he
> loves my kid. How can I entrust our lives, and give up my apartment,
> and go live there?" I was feeling really miserable, and I was driving
> home and this psychologist cut me off on Chain Bridge Road and she
> starts honking me down, tells me to pull over, and sticks her head
> out the window, holding up traffic the whole time. "Don't blow it!"
> she goes. "This guy's good. Don't be an *asshole*!" And she gives me
> her card, and says, "Good luck," and she drives off.

Laurie made up her mind; she and Tom have now been married
for nine years. He's been a generous stepfather to Sean, who now
has two younger half-brothers. The marriage is affectionate and
stable, and Laurie's top priorities, honesty and fidelity, are honored.
The marriage has also had its share of disappointments, about
which Laurie is clear-headed and candid. "I think that in marriage
you choose your role," she says. Tom's a workaholic, and after
years of trying to drag him away from his desk, she finally pulled
back. "Now I say, 'If you really think this shitty business is worth
killing yourself over, be my guest,'" Laurie says bluntly. "I've
learned to tend my emotional needs more, not look for them to be
met by Tom." Is that a terrible loss or a sensible accommodation?
It's a judgment call. The compromises are worth it to Laurie, and
she is happy with her life.

Dina, the immigration attorney from Sacramento, would have
stayed single a little longer, but Rodney, her second husband, made
it clear it was now or never. Six years later the marriage is going
strong. What most attracted Dina to Rodney was how he behaved
with her two kids. "It was exactly what I had wanted Alan to be.
He was fun, he was loving, he played with them, they loved him.
He said right off, 'I understand it's a package of three.'" Dina
prefers being married to being single, because "it gives me an identity that I like," but adds, "I don't think I behave differently." What
makes this relationship different is that "we talk a lot more. I was
not good at all at expressing my feelings with Alan. I'm willing to
work at this one, to make it work."

Women Remarry to Connect with Someone Who Shares Their Basic Values

"It dawned on me that we could stand back to back and see the world in the same way."

"I didn't think I would ever get married again, because I didn't think anybody would want a single mother with a child," says Caroline. Four years later she met a man at a Thanksgiving Day party. Gavin was smart, affectionate, and supportive. Both English expatriates, they married ten months later. Much of Gavin's appeal lay in the fact that, unlike her first husband, he was as ambitious as Caroline if not more so, and he wanted the same things out of life that she did. That thought is what sustained Caroline during a difficult period two years into the marriage, when they almost split up. "We were driving each other nuts," recalls Caroline, "*but* it dawned on me at the time that we could stand back to back and look at the world and see it in the same way. That was the thing that made me say, 'We can work it out,' despite the toothpaste and the snoring and the whatever it is."

Another plus was Gavin's intuitive understanding of the fact that Caroline's eight-year-old daughter was close to her own father and didn't need a second one. Instead he forged a separate and distinct relationship with Heather. "They found their own common ground. Oftentimes they ganged up against me," says Caroline with a smile. "They get along very well, they're very close."

Susannah's second husband is a sensitive, touchy-feely kind of guy, while she's a cut-and-dried rationalist, but this time she was looking for more than surface compatibility. "We're very different people, we knew that from the start," says Susannah about Clifford, a coffee importer who takes pride in his wife's standing as a well-known film archivist. "But fundamental values, and feelings about family and kids, are very aligned." Susannah remarried in large part because she wanted another child, and family matters a great deal to her. "If I really look at it, I think I always understood being a mom but I don't think I've ever understood being a wife," she says candidly. "There's a lot of things that I don't love about marriage, a whole bunch of sharing that I'm not good about. I don't always like having to talk over certain decisions about my kids." Their son Ezra is now twelve, and their marriage appears to be an unusually strong and supportive one. This time around Susannah understands what it's based on.

> Remember when I told you my parents had a great marriage and I didn't think it was a big deal? What I found out was that it's a big deal. It isn't just enough to find out that you like the same movies and walk nicely together down the street, that you both get on. Those comfortable things may cover up some fundamental incompatibilities. Now I understand how important it is to honor the fundamentals in order to get through the tough stuff, or even just the boring stuff, like who's doing the laundry.

That doesn't mean Susannah didn't enjoy being single. "I was twenty-eight years old [when I got divorced], and one of the peak moments of my life was my thirtieth birthday. I thought thirty sounded very womanly, very sophisticated. I was on my own." Remarried at thirty-five, seven years after her divorce, she's glad that she had those years of being alone. "It certainly makes getting older easier," she notes. "I think it must be frightening for somebody who's been with somebody their whole life."

"Mine is the way marriage is supposed to be. Give and take."

When the fundamental values are shared in a romantic relationship, so, it seems, is mutual respect. A greater overall compatibility in how two people see things and what they like to do means that the man can encourage independence and growth. Separate goals are honored. Opinions are exchanged instead of lines being drawn. "Mine is the way marriage is supposed to be," says Sophia, who spent ten years in a profoundly abusive first marriage. "Give and take. Respect for one another. It just flowed so smoothly."

WENDY
Take your time. The textbook case is the disastrous second marriage. Everybody I know just thinks they have to get married again so quickly, it's like their worth depends on them being a married woman.

Women Remarry for Romantic Companionship

"I think that making promises and living life with someone is a gift."

An abusive marriage can drum the very idea of romance out of a woman's head. She may close down in self-protection, until maturity, affection, and circumstance combine to coax the idea of romance back into the realm of possibility. Though Amanda, an architect, was only twenty-six when she divorced, she thought she'd probably be alone for the rest of her life. "I didn't think I could actually make it work, actually live with somebody. I questioned myself, thought, 'My God, maybe I'm just really a selfish person, I don't know how to share.'" She eventually realized otherwise, though in the interim she retreated into a very private life, working, painting, and raising her young son. Ten years have passed; two years ago Amanda married a man she had known for several years before the romance developed, and describes as her best friend.

No longer the shy grad student who needed a partner to be the smart, charismatic, funny one, she went into this marriage feeling "like I could take care of myself. I don't need anyone to support me." She embraces what marriage has to offer, as she makes movingly clear. "I think that making promises and living life with someone, where you're kind of witnessing each others' lives, is a gift, a tremendous opportunity," she explains. "You're a witness for them, and you can see things, create a history together, and help each other." Amanda also feels that being married helps to define the nature of the relationship, though not the commitment aspect of it, citing gay couples she knows whose relationships are as solid and monogamous as any legally sanctioned one.

Hillary also found romance without losing herself; her portrait of her marriage to Matthew is one of the loveliest I've ever encountered. Hillary calls him "the man who came for breakfast and never left." They lived together for a year and married on the anniversary of the day he moved in so they'd have all their anniversaries in the same place, thirteen so far. They have eleven-year-old twins. "This time around I was more conscious of the fact that what you see is what you get; it's not going to change. So I did think about 'Is this something I'm willing to live with?'" says Hillary. "I think it has a whole lot to do with trust. When I met Matthew something inside of me relaxed which has never relaxed all my life. I know that it has a lot to do with the right timing, the right place in your life. But I just knew he would never let me down on the important stuff, and he never has."

In therapy Hillary came to understand that the first time around she had married someone who disapproved of her because she didn't approve of herself. She came to her second marriage a very different person, confident and happy with herself. "I don't need to have this; I don't *need* Matthew," she points out. Hillary thinks the idea that everything between husband and wife should be shared, from bank accounts to free time, is "all malarkey. If you have this soulmate, nothing has to be joint because you are joined. We have joint nothing. I had all that in the first marriage, but I never had the soulmate."

Hillary has pinpointed the two paradoxical poles of a successful relationship: independence and interdependence. Sharing is crucial, and so is generosity, but privacy and autonomy are important too. Each partner's strong sense of self is integral to the relationship, so that the balance of power can go back and forth. Another essential ingredient is a good sex life, which requires care and feeding. Hillary and Matthew make sure to get a baby-sitter for the twins every so often and check into a hotel room, and she usually packs something slinky to surprise him with.

Why Women Haven't Remarried

Divorce makes the mundane and unattractive aspects of marriage glaringly apparent, and the memory of those drawbacks doesn't always fade away as time passes. Many women who had endured bad marriages were not interested in being subject to a husband's moods and whims again, and were unable or unwilling to conceive of marriage in other terms. For other women, achieving a satisfactory standard of living was itself enough to squelch the desire to remarry. Another key deterrent was psychological: they had learned, often painfully, to assert themselves and take control of their lives, and they weren't interested in relinquishing that autonomy. Traumatized by her divorce and aware that second marriages are even less likely to last, Rachel, the actress, readily admits that the prospect "petrifies" her. "I don't want to be divorced twice," she explains.

Why Bother?

"Marry again? Absolutely not interested."

Burned by a bad experience, many women feel the downside of marriage outweighs the benefits. Forsaking a shared history and the comfort of lifetime companionship is indeed painful, but so is recuperating from the shattered illusion that marriage offers surefire shelter from the storms of life. The compromises necessary to make marriage work seem too great for many divorced women, for whom once around is enough.

Divorce is a great reminder that for much of history marriage arrangements revolved around property. The fights over who gets what and the adjustments to a lower standard of living make many women question the upside of signing a marriage contract again, especially when they have made good lives on their own. In the twentieth century, for better or worse, we have come to demand romance and personal fulfillment of matrimony, along with financial security, a tall order at the best of times. We have grown less tolerant of unhappiness in marriage, and less afraid of divorce.

For years Violet was brutalized by her marriage, but when she was in it she was just as terrified by the alternative: being alone in the world without the shelter of matrimony. Now free of the illusion that marriage protected her, a message reinforced daily by her job at a battered women's shelter, Violet doesn't miss the daily reality either. Now fifty-three, she has been single for over twenty years. "Before, I thought I was missing something because I didn't have a mate," she says, "but now my friends that I thought were so happy call me on the phone for hours, crying." Deeply and happily focused on her work with children, Violet says matter-of-factly, "I've gotten used to being alone."

Both her age and her eminently satisfactory living arrangement rule out remarriage for Olivia. "My feeling is that if I were to get married at my age it would be to take care of a sick old man," she says frankly, "and I don't want to do it." Married for forty-one of her sixty-seven years (a first marriage at nineteen lasted six years),

Olivia has completely and contentedly closed the door on any Prince Charmings. Over lunch not long ago an old friend asked her whether she was interested in marrying again. "I said, 'Absolutely not interested.' He said, 'Well, wait a minute. What if some really nice guy like me came along . . . ?' And then he starts sending me valentines, he was so cute, it was a little plea from him not to give up on the whole thing. I guess I was breaking the tradition there too," acknowledges Olivia perfectly happily. "I as a woman was supposed to be hanging around waiting for a man to complete myself. I'm really not interested." She shares a house with an old friend who was widowed several years after Olivia's divorce. "Oh, it's *lovely*," she says, beaming. "I can't *tell* you how nice. You're equal, you don't have any of this societal role stuff." Olivia says she wouldn't go back to living with a man for anything in the world. "I've got the best of both worlds. I'm single but I'm not alone, and that's pretty damn great."

Women Get What They Need Outside of Marriage

"If I lived with him I think the sex would become dutiful, and that's a killer."

In her early forties when her marriage started coming apart, Wendy remembers making her peace with the possibility that there might never be another man in her life, "but there have been a lot of men in my life since then," she says. "Sex was great, it was wonderful." Her brother-in-law's reaction to her single state—"You're a good-looking woman, I don't understand why you don't get married"—was typical, but Wendy shakes her head dismissively. "Maybe I'm afraid that I'd become that doormat wife again."

Although happy with her independent life and career as a food writer and cookbook author, she did briefly consider remarrying her ex-husband, two years after the divorce. "We were in a restaurant and he said, 'You know, Wendy, it's about time we stopped this nonsense and got this marriage back together again.' I think if he'd done it in a sweet way, or said, 'I've really learned a lot, do you think we could try again?' I might have said yes," admits Wendy, "but he sort of made it a decree." Now fifty-seven, she acknowledges that the older she gets the less likely she is to remarry, "but that's all right."

Maybe if I didn't have a good relationship with these children and my grandchildren and know that I will always have family, that there will always be other human beings that I can touch in a physical way [it would be different]. That gives me strength. I had lunch today with a friend who's never been married, her parents are elderly, and there's nobody she can touch. I think, "What does the future hold for her?" and I think the future's fine for me.

For a number of years Wendy has been involved with a Brazilian businessman. He lives in São Paulo, so they see each other only a few times a year. "When we get too old for flying or one of us loses our health and the relationship ends, it will be very sad for me, because I can go and feel sexy and wear a black nightie and still look good in it, have that part of my life," says Wendy, "but I'll tell you, if I lived with him I think the sex would become dutiful, and that's a killer." Wendy's social and sensual needs are being met just fine, and she feels marriage would be detrimental to her sexual ones.

Remarriage Isn't Always in Women's Best Interests

"An intelligent woman who can take care of herself has better things to do than fold a man's socks into little balls."

Anneke, the romance writer from Tucson, particularly enjoys doing what she feels like doing without having to account to anyone. "I don't think I'd want to get married again," she declares. Her husband wanted her to spend 100 percent of her leisure time with him, "so it was a problem if I wanted to go out to a movie or go to dinner with a girlfriend or take an evening class," she explains. "Now if I feel like going out every night, I can, and I don't have to make excuses or anything." At this point, a year and a half after her divorce, Anneke can't imagine someone coming along who could persuade her to forsake these distinct pleasures. Almost fifty, she enjoys her freedom; childless, she isn't bogged down by maternal obligations. "I would like to have a committed relationship, but I'm not really sure I want to live with anybody again," she says firmly. "I think that unless a woman has a really good relationship with a man, she's better off alone. Because an intelligent woman who can take care of herself and support herself is better off doing her own thing and choosing what she wants to do than folding a man's socks into little balls."

Forty-seven and also childless, Natasha is very serious about her new boyfriend, but she's afraid that remarriage would erode her personal growth and the quality of the relationship. She figures it took her two years "to really realize that I was separated and divorced," and that's when she met Mark, a restaurateur. The romance sailed along until Natasha's divorce papers arrived in the mail and she pulled back in a big way. Finally, at Mark's gentle insistence, Natasha confessed to feeling really panicked, "getting this whole catch-in-the-chest feeling again." (Celeste too nearly broke up with her now-fiancé when her papers came through. Describing her behavior as "lunatic," Celeste explains, "what it was was *terror*. I was finally really available and it's different, in some profound way.")

Referring to the extroverted self she's regained in the three years since her divorce, Natasha observes that "maybe that's what I'm fearing in a way, you know, the loss of this part of myself that I really enjoy and want to hold on to." Mark gave Natasha plenty of space and the relationship has gotten back on track. She'd like to try living with him, and they've talked about marriage, "but it really brings up a lot of issues," she admits with a nervous laugh. "Marriage changes things. I really understand more and more that it's an institution, and that it works if you are willing to be institutionalized." She thinks success depends on constant vigilance about two things: being too dependent ("you're just going to have to work chipping away at the codependency"), and the tendency to let one's separate identity slip away. "Sexually, socially, in every way, it's work, as a woman and as an individual," comments Natasha. "I'm always checking where I am in the relationship."

Committed Relationships Outside of Marriage

Living together without getting married is increasingly popular in the United States. Nearly 3.7 million American households are made up of unmarried couples, a number that has climbed 85 percent over the past decade and increased sevenfold since 1970. One pragmatic reason is that unmarried dual-income couples across the economic spectrum have a considerable tax advantage over their married counterparts, and older couples are a big part of the boom. Cohabitors, however, are at a distinct disadvantage when it comes to tenants' rights, co-insurance, inheritance, and custody issues, which is why Eileen completely rules out the possibility of giving up her own place to move in with someone. "I've had a lot of girl-friends who put themselves into relationships and all of a sudden it breaks up, and it was his place, and they're kind of like, 'Oh, what do I do now?' They haven't got a leg to stand on. Not for me."

The absence of this mutual financial and legal obligation, though, is precisely what makes living together preferable to getting married for many couples: the arrangement is less defined by social convention. Living together offers couples the opportunity to work out who is to be responsible for what in a conscious way, rather than settling into the established roles of husband and wife. Free of such preconceived notions, the domestic landscape becomes a blank slate on which a new and more mutually satisfying script can be written as life together evolves.

In their study of thousands of American couples and what made them work, social scientists Philip Blumstein and Pepper Schwartz felt that what cohabiting partners gave and received was particularly well balanced. They noted that for women the arrangement involved choosing autonomy over the security of being supported, and required that men put establishing their own economic identities over the need to control their partners. "This should make for a successful financial partnership as long as the woman is *really* not going to resent the man for rejecting the provider role, and as long as the man is *really* happy to live with an independent woman who wants an equal say in the relationship," they wrote. Blumstein and Schwartz went on to observe that "since these couples are going against the established traditions for men and for women . . . they face an intriguing social challenge." That's for sure.

Although such partnerships are extremely common, they lack the social sanction of marriage, a fact underscored by the absence of words to describe the players. "Lover" is smarmy, "significant other" impossible to say with a straight face, "beau" too precious, and when my children refer to "Mom's boyfriend," I cringe. He's no boy, and we're much more than friends. When tending to joint-family logistics, I refer to myself as Bob's "wife-substitute" (sounds like some kind of health food supplement). Bob's pick is "current lifetime companion," which, to my chagrin, never fails to get a rise out of me. During the divorce proceedings he referred to his not-yet-ex-wife as his "non-wife," and she remains far more his wife in both our minds than I, though perhaps that will be changed by the weight of years. We're in it for the long haul but we don't even live together, which makes our social status even more ambiguous.

Maintaining a relationship outside of marriage means that there are no roles to fall into, and none to fall back on either. For someone (like me) with a low tolerance for ambiguity, it can be exhausting, but it keeps me on my toes, and I wouldn't swap it. I've already got the kids, the sofa, and the mortgage, and at this point remarriage holds all the charms of a sharp stick in the eye. To me the sanction of the state is unimportant; a piece of paper no more guarantees permanence, God knows, than a wedding ring guarantees fidelity. Furthermore, I think being married would make it easier to take each other for granted, the most unwelcome prospect of all.

Some day Bob and I will promise ourselves to each other in front of friends and family, but I doubt that rings and marriage licenses will figure in. (Again, vocabulary falls short: "commitment ceremony" conjures up an image of men in white waiting with straitjackets.) We do want to live together, but four other people figure into the equation—his two children live with him—and the idea of the Brady Bunch makes the children's collective hair stand on end. Right now their needs come first. Besides, it's very romantic this way. There's a great solution, if only we could afford it: separate places, side by side. "How many kitchens?" asked his mother at her fiftieth anniversary dinner this May. Her face fell, God bless her, when we each shot two fingers into the air.

"I don't think paper makes people have committed relationships."

For ten years Tory was involved with a man she wanted to marry, but when that relationship ended so did her interest in matrimony. Now in her late forties, Tory's been single for over twenty years and can't count the number of times people have said to her, "'I'm surprised you never got married again. How come you're still single?' Like, 'Didn't anyone want to marry you?'" she elaborates with an exasperated toss of her head, imitating their pitying tone. She resents the question, and feels that being unmarried indicates "that you're a whole person, that [marriage] isn't a requirement. I like the independence. I enjoy the freedom." For the last three years Tory's been very involved with "someone who's got a very high sense of autonomy like I do, and it's comfortable. But it's not easy to find somebody like that," she readily concedes. She and Donald live together but she travels a lot and they give each other plenty of space. In the absence of any pressing legal or financial issues, Tory feels no desire to tie the knot. "I don't think paper makes people have committed relationships," she says firmly.

"We have to write our own scripts."

So how can marriage be adjusted to deliver what it promises— happily-ever-after? We can start by learning some valuable lessons from homosexual relationships, where conventional habits and hierarchies don't necessarily apply. For forty years Rosemary, the air-traffic controller, lived her life as a straight woman; she has been a lesbian for six. Until she got to know more gay people, Rosemary shared the perception that in male and female gay relationships, one person has to be masculine and the other feminine, "butch and nelly if you will," but now she declares it not so.

I think what's really different about homosexual relationships nowadays is that we don't have any real rules. In general in heterosexual relationships one person takes the dominant role. It's not always the man, though most frequently it is. What I'm seeing in gay relationships is that there's a lot more switching of roles, taking turns, because nobody is the only owner of a penis. Nobody feels obliged to be always the guy. So we have to write our own scripts.

Now forty-six, she has had three serious lesbian relationships, including a deeply committed one with the woman she lives with, which they expect will be permanent. When issues come up, it's pretty interesting, because "one of us will say something and the other will say, 'Well, yeah, but that's how straight people do it. We don't *have* to do it that way unless we want to.'"

Imagine a truly democratic, gender-blind household, where responsibilities and resources are perennially up for grabs. Imagine allocating them according to present ability and need, rather than relying unthinkingly on patterns of deferral and dominance. Imagine how demanding it would be to navigate without the signposts of gender roles, and how different and exhilarating it would be to call such a place home. It might be inhabited by a husband and wife, or marriage might turn out to be less relevant than we can presently imagine.

Making Relationships Work Better

Many happy and productive unions prove marriage to be eminently worthwhile, for wives as well as husbands, when both partners value each other's contributions equally. In a 1995 poll of four thousand readers, *New Woman* magazine found that women in Egalitarian Marriages, defined as those in which both spouses probably work full-time and share domestic and child-rearing chores more or less equally, are by far the happiest. The alternatives were Traditional Marriage, in which the husband is the principal breadwinner and the wife takes care of the home and the children, and In-betweens, in which the wives described the allocation of responsibilities as "something in-between" the egalitarian arrangement and the traditional one. ("Isn't there a category where the wives are the boss?" asked my daughter.) In summary, Stephanie von Hirschberg, senior editor and the author of the article, writes that shared power and responsibility "seems to be *crucial* to a woman's happiness in marriage."

Three other elements also increase the odds of a successful long-term relationship: chemistry, the desire to care for each other, and common life goals. Anybody can live with anybody, depending on how many compromises he or she is willing to make. This is an essential part of any relationship, heaven knows, and only the person making the compromises can decide whether they're worth it in the long term. The arrangement tends to be worthwhile as long as the give-and-take balances out in the long run. It may be 80 percent give and 20 percent take one day, or for a month, or even for a year or two; then it needs to be 20–80 for a while. Fluidity is the key, because circumstances change: someone goes back to school; a parent moves in or a child moves out; somebody gets sick; one person hangs on to a paycheck while the other takes a professional flyer. Forget roles; it doesn't matter who does what. It's fine to be a stay-at-home mom while the kids are small, and it's also fine to hire a baby-sitter and head for the office. It doesn't matter if she washes every dish as long as he assumes some other task, domestic or otherwise, that she dislikes. All solutions are personal.

Men and women need to treat each other as equals, which is no mean feat in a world that does not. For husbands and wives the effort requires imagination and the courage to be unconventional, because the structure of traditional marriage is not egalitarian. Wives want their careers to be as important as their mates'; wage parity will bring this about eventually, but until then, it's a tall order. Wives want spouses who really listen and who genuinely care about their opinions. They want to respect and care about their husbands, and to be loved and respected in return.

Instead of giving in to their husbands' desires as a matter of course, or expecting husbands to magically intuit their desires, wives need to articulate what they want. This means turning a deaf ear to inner and outer voices that say such behavior is selfish or bitchy. It means learning and daring to express feelings both loving and murderous. Many women interviewed for this book said that poor communication lay at the heart of their marital problems, and that expressing themselves was the key to a better relationship. Although it wasn't a conscious goal at the beginning of her second marriage, Laurie makes "trying to keep lines of communication open" a top priority. "It's so easy to get lost," she observes. In my case, I spent much of my marriage afraid that my husband was angry at me, and thoroughly repressing my anger at him. Now I ask Bob directly, "Are you mad at me?" "Huh?" is his usual response. And when I'm mad, I take a deep breath—I still have to steel myself—and say what's bothering me. The habit of censoring oneself dies hard, but as these women have learned, it is poisonous, not only to their relationships but to their own happiness.

What the Future Holds

Will women wise up and start asking less of marriage? No, and it's just as well: women should ask more of relationships, not less.

Marriage isn't the solution, or the problem. Whether the institution is redefined or abandoned is really not the point, nor does it matter whether people pair up as husband and wife or opt for a wedding-free zone. Everything is up for grabs between conscious adults, which is both frightening and heartening, and wedlock guarantees nothing. Marriage itself is becoming less relevant.

Blaming men for the sorry state of gender relations is as easy—and unconstructive—as blaming marriage for the unfulfilling nature of many unions. The point is not that husbands are brutal and wives hapless, or women righteous and men inadequate; if men are self-centered, it is in large part because women are socialized to let them get away with it. Asking less of men is one way to reconcile differing expectations, but one that demeans both men and women. Because we are stuck with the very human tendency to get away with whatever we can, to settle the battle between the sexes we have to revamp the social contract. This is especially difficult when so many of our religious and political leaders are looking backwards toward outmoded models rather than ahead toward innovative solutions.

The fact that marriage is oppressive for so many women is not the fault of the institution or the players but of society. When men have more power than women, women are readily exploitable. I got divorced because I felt personally, not politically, oppressed. But gradually I saw that what doomed my marriage was less my ex-husband's and my personal failings, many though they were, than social forces about which I was completely clueless at the time.

To stop being stifled, I and the other women in this book chose to end our marriages. Avoiding marriage altogether is another way to sidestep the imbalance of power between husbands and wives. So is working with a like-minded man to craft a truly egalitarian marriage. All three solutions are unconventional and all have political repercussions, but the last is the greatest challenge of all because it calls for social change on an individual and collective level. Until men and women are paid the same wages and are jointly responsible for family life, there will be no such thing as equal opportunity. Until men and women come to the table—be it in the kitchen or the boardroom—as true peers, the egalitarian marriage will remain the exception, if not a downright contradiction of terms.

Improving the odds of happy marriages means working to end

wage and job discrimination against women. It means providing economic and social support for single mothers, and for men who want to participate fully in raising children. Women who point out inequities worry about being accused of bashing men, but as Naomi Wolf says, you can hate sexism without hating men. The world is a sexist place, but if each of us responds to it in a nonsexist way, society will accommodate us because it must: individual action sparks social change. When a woman ends an oppressive marriage, it is a creative act of resilience and survival; she is not deviant or immoral, but a brave pathfinder. Until we figure out a better way to do marriage, divorce is here to stay.

It is women, not men, who lose sight of the underlying truth grasped by my daughter after a bedtime schmooze about gender politics. "Mom," she said, "women are awesome." We have nothing to lose and much to gain: the chance to reinvent partnership and show our children what love between equals can be all about. Slowly, by trial and error, each of us is writing her own script. It's part of a collective one that's stacking up, page by page, as women make their way to financial, social, and personal equality with men.

WHAT SINGLE PIECE OF ADVICE WOULD YOU GIVE TO A WOMAN WHO IS CONTEMPLATING ENDING HER MARRIAGE?

SOPHIA

Don't be afraid to be in a column that says you're a single parent. Don't be afraid to be in a column that says you're a victim of spouse abuse. Get to know yourself more, because there are a lot of hidden qualities you've been stripped of that you didn't even know were there.

BEVERLEY

Make a list of everything that pertains to your marriage, good and bad, and then you can decide whether these problems you're having are really real and big. If that person is willing to make an effort to work it out, you really should do that. If they aren't, and you've had enough, then you need to make a decision and stick to it, that's the biggest thing. All that wavering back and forth is a mess. As painful as things can be, there are times that you have to let go.

HANNAH

A lot of women know that this is what they need to do, but they won't do it because it's so hard, [and] it *is* going to be hard. But you come out on that other side and it's worth it.

YOLANDA

My first impulse, based on my experience, would be to say, "Do it, just *do* it." My second impulse—no, my second impulse would be to say, "Do it, just do it." And if it's wrong, that's not incurable either.

CELESTE

Be brave. I think that perpetual singlehood is a possibility, and I can't say that it's ideal. But I don't think that it's as bad as living in a marriage that's not working. That's just misery.

RACHEL

I think that a marriage, a person you're in love with, should encourage the best parts of you to rise to the top, like cream. And if you sort of suppress them, or if you feel at all held back, then you should get out as fast as possible.

TORY

When I talk to women who are apprehensive about getting out of a marriage, their biggest concern is "Can I take care of myself. Will I be able to earn a living? Will I be able to get out, to be able to still go to tennis, and provide my children with singing lessons?" . . . Get a part-time job. Start proving your ability to work, and test your capabilities. You'll probably be surprised at how much you can accomplish compared to Susie Q down the street who just got out of school. Experience counts a lot.

LORRAINE

I'm proud that I *did* it, because my life is my own all over again. It's my life. And my kids are fine. You can tell anybody that's thinking about it if they've struggled, and they really know it's the right thing to do, do it.

FRANCINE

Get a reputable therapist. No one else can help you in that way that I think you need.

AMY

Two things. One is, talk to people. Talk to women who have done it, because you get something very positive out of it. I also think that people should try living apart before they get to the point of terrible fighting and calling in the lawyers. Live apart. See how it feels.

MEGAN

If you're not happy, you are the only one in life that can make it happen. Nobody else, not your spouse, not your kids, not your mother and father, brothers and sisters, not your job. Only you. And you've got to find out in life what's going to make you happy.

NATASHA

That little voice is really never wrong. You may not think so at the time, but trust that little voice.

GINNY

If you look to your spouse for every source of happiness, you're set up for failure big time, because ultimately you are responsible for yourself and your spouse isn't. I think most people contemplate divorce because they think that their spouse is the cause of their problems. So if you're leaving him because you think that something else is brighter on the other side, think twice.

KEIKO

Despite this constant panic over financial independence, I've never been more hopeful in my life. I would say to someone who's miserable in a marriage, "Life is very short and there are wonderful things to be had. If you want those things for yourself, a chance at something better, you have to take some risks."

LAURIE

Go on with your life. Give yourself a chance to be happy. And don't marry the same schmuck again.

Index
of Women
Interviewed

Source Notes

Introduction

PAGE

xxii *the chances of a woman over thirty* An exercise in limited demographics (based entirely on census data, the study recognizes only two categories of women, single and married, so no account is made for live-togethers, lesbians, and alone-by-choicers; only the percentages of finding an older mate were analyzed; no college-educated single women were interviewed at all), another piece of its bottom line is the projection that eight out of ten white college-educated women will marry, a decrease of only 10 percent from preceding generations. It didn't mention the fact that both the percentage of those who "never married" and the median age of marriage were almost identical in 1890. Susan Faludi, *Backlash* (New York: Voyager, 1992), pp. 59–63.

xxii *Well-publicized figures showed a divorced man's income soaring* Lenore Weitzman, *The Divorce Revolution: The Unexpected Social Consequences for Women and Children in America* (New York: Free Press, 1985), p. xii. See first two Notes to page 98.

xxii *wives have always sought divorce in greater numbers* This is confirmed by a wide variety of sources:

- Seventy-five percent of divorces are initiated by women. Augustus Y. Napier, *The Fragile Bond: In Search of an Equal, Intimate, and Enduring Marriage* (New York: HarperPerennial, 1990), p. 338.
- Approximately 61 percent of all divorces in 1988, the latest date for which such statistics were available, were petitioned by the

wife. National Center for Health Statistics, in the "Monthly Vital Statistics Report" 39(12), Suppl. 2, 21 May 1991.

- Between two-thirds and three-quarters of divorces are initiated by women. Constance Ahrons, *The Good Divorce* (New York: HarperPerennial, 1995), p. 35. In three-quarters of the families studied by the California Children of Divorce Project, wives took the first step. Judith S. Wallerstein and Joan Berlin Kelly, *Surviving the Breakup: How Children and Parents Cope with Divorce* (New York: Basic Books, 1980), p. 17.
- See also next-to-last Note to page 63.

xxv *But even Lenore Weitzman* Elizabeth Gleick, "Should This Marriage Be Saved?" *Time*, 27 February 1995, p. 53.

xxv *"Even the longer-married older housewives"* Weitzman, *The Divorce Revolution*, p. 346.

xxvi *Jerry Falwell, head of the Moral Majority, declared* William H. Chafe, *Paradox of Change* (New York: Oxford University Press, 1991), p. 219.

xxvi *The characterization of divorced people* Ahrons, *The Good Divorce*, p. 12.

xxvi *"I have come to believe that when the end comes"* Mary Catherine Bateson, *Composing a Life* (New York: Plume, 1990), p. 208.

Chapter 1 • Shattering the Illusion

1 *. . . If I continue to endure you* William Congreve, *The Way of the World* in *The Comedies of William Congreve* (New York: Penguin Books, 1985), p. 379.

1 *Mercedes, the current Mrs. Sid Bass, has metamorphosed* "Page Six," *New York Post*, 15 August 1994.

3 *In fact, wives constitute the most depressed segment* A study conducted by epidemiologist Martha L. Bruce and psychiatrist Kathleen M. Kim, of the Yale University School of Medicine, found that "happily married women suffered nearly four times as much severe depression as happily married men" (*Science News*, 18 July 1992). Reviewing numerous studies, Maggie Scarf found that the number of females diagnosed with depression was two to six times higher than the number of men. Maggie Scarf, *Unfinished Business: Pressure Points in the Lives of Women* (New York: Ballantine Books, 1980), p. 3.

7 *"Women who 'balanced' felt 'too powerful'"* Arlie Hochschild, *The Second Shift* (New York: Avon, 1989), p. 222.

8 *Men do more around the house than they used to* Kathleen Hall Jamieson, *Beyond the Double Bind: Women and Leadership* (New York: Oxford University Press, 1995), p. 62.

9 *"I have yet to meet a woman"* Anne Wilson Schaef, *Women's Reality* (New York: HarperCollins, 1992), p. 75.

9 *"our language is laden with unwed mothers"* Jamieson, *Beyond the Double Bind*, p. 194.

12 *"If she can annihilate herself altogether"* Dalma Heyn, *The Erotic Silence of the American Wife* (New York: Signet, 1993), p. 65.

13 *O. J. Simpson described himself* "Simpson Describes Self as 'Controlling,'" Associated Press, *San Francisco Chronicle*, 3 February 1996.

17 *she puts her "feeling self on ice"* Dana Crowley Jack, *Silencing the Self: Women and Depression* (New York: HarperPerennial, 1993), p. 137.

18 See Deborah Tannen, *You Just Don't Understand* (New York: William Morrow, 1990).

Chapter 2 • Hard Choices
25 *a woman over thirty is less likely to get married than to be shot by a terrorist* Faludi, *Backlash*, p. 41; also see first note.

25 *"But this quiescence, from the point of view of outward behavior"* Scarf, *Unfinished Business*, p. 94.

26 *"The terror is that if she responds to her own needs and wants"* Heyn, *The Erotic Silence of the American Wife*, p. xiv.

27 *The current consensus is that divorced women* Hanna Rosin, "Separation Anxiety," *New Republic*, 6 May 1996, p. 16.

27 *One-third of families headed by women* Sara E. Rix, ed., *The American Woman 1990–91* (New York: Norton, 1990), as quoted in *WAC Stats: The Facts About Women*, Women's Action Coalition (New York: New Press, 1993), p. 41.

33 *"When you . . . make divorce quick and easy"* cited by Rosin, "Separation Anxiety," p. 14.

35 *"I'm the one who did it, I divorced"* Ann Patchett, "The Sacrament of Divorce," in *Women on Divorce: A Bedside Companion*, ed. Penny Kaganoff and Susan Spano (New York: Harcourt Brace, 1995), p. 9.

36 *"In the Retrouvaille world no marriage is too broken"* Rosin, "Separation Anxiety," p. 17.

40 *Research shows that people find treatment* Consumer Reports, November 1995.

42 *"'women's intuition' . . . may be nothing more"* Susan Brownmiller, *Femininity* (New York: Fawcett Columbine, 1984), p. 197.

49 *As many as one in seven women have been raped by their husbands* Pollitt, *Reasonable Creatures* (New York: Alfred A. Knopf, 1994), p. 6.

49 *Every fifteen seconds a woman is battered* Statistics compiled by the National Woman Abuse Prevention Project, http://users.mdn.net/mcfoc/shelter.htm.

Chapter 3 • Laws and Lawyers
55 *compared to 14 percent of divorced women nationwide* Faludi, *Backlash*, p. 95.

55 *So much for the stereotype* See fourth Note to page 98.

56 *48 percent of married women . . . bring in half or more of their household income* Tamar Lewin, "Women Are Becoming Equal Providers," *New York Times*, 11 May 1995.

62 *In colonial Massachusetts penalties for violating the marriage vows* Glenda Riley, *Divorce: An American Tradition* (New York: Oxford University Press, 1991), p. 15.

63   *the number of divorces grew steadily*   Roderick Phillips, *Untying the Knot: A Short History of Divorce*, (Cambridge and New York: Cambridge University Press, 1991), p. 235 (graph).

63   *In 1868 women's suffrage leader Elizabeth Cady Stanton declared*   The quote is from *The Revolution*, a women's rights journal edited by Stanton, quoted in Riley, *Divorce*, p. 84.

63   *Laws were liberalized to recognize a broader range of grounds for divorce*   Divorce historian Roderick Phillips observes that divorce law liberalization at mid-century was not intended to emancipate women or to make divorce more available to the lower classes. He writes, "The grounds recognized . . . were assumed to be male offenses for the most part, such that divorce was less a way of freeing women than of protecting them. To this extent the divorce laws were part of a complex of paternalistic legislation that sought to protect women from the most harmful implications of their inferior status without attempting to change their status significantly." Phillips, *Untying the Knot*, p. 172.

63   *State legislatures passed more than one hundred restrictive laws*   Faludi, *Backlash*, p. 162.

63   *more than two-thirds of all divorces*   The percentage of female initiators at the turn of the century was lower in other Western countries than in the United States, but, observes Phillips, "it was usually because divorce laws discriminated against women . . . [When this ended,] petitions by women quickly rose to form a majority." Phillips, *Untying the Knot*, p. 228.

63   *Many divorcing women were mothers*   The Census Bureau reports that the number of divorced single mothers is currently rising about 2 percent a year, compared with 9 percent annually in the 1970s. Barbara Vobejeda, "Divorce Rates Slowing," *Washington Post*, 24 June 1993, p. A21.

64   *nine states, with California paving the way, passed community property provisions*   The other states are Arizona, Idaho, Louisiana, Nevada, New Mexico, Texas, Washington, and Wisconsin. The rest have equitable distribution laws. Karen Winner, *Divorced from Justice: The Abuse of Women by Divorce Lawyers and Judges* (New York: ReganBooks, 1996) p. 42.

64   *But because lawyers and judges*   Winner, *Divorced from Justice*, p. 53.

65   *"Most evidence suggests that no-fault had little effect"*   John Leland, "Tightening the Knot," *Newsweek*, 19 February 1996, p. 73.

65   *"If anything, the divorce law reforms were a response"*   Phillips, *Untying the Knot*, p. 243.

65   *The American divorce rate has in fact declined slightly in recent years*   A high of 5.2 people per 1,000 in 1980 has declined to 4.8 in 1990 and 4.6 in 1994. Cynthia Costello and Barbara Kivimae Krimgold, eds., *The American Woman 1996–1997—Women and Work* (New York: Norton, 1996), p. 360.

65   *Nixing no-fault is hardly likely to encourage the marriage-shy*   Rosin, "Separation Anxiety," p. 14.

65 *Horace Greeley, for example, complained that too many people* Riley, *Divorce,* p. 84.

65 *Measures currently before numerous state legislatures* The states are Georgia, Idaho, Illinois, Iowa, Michigan, Minnesota, Pennsylvania, Virginia, and Washington. Dirk Johnson, "Finding Fault with No Fault Divorce," *International Herald Tribune,* 13 February 1996, p. 1.

65 *A front-page article in the* International Herald Tribune Johnson, "Finding Fault with No Fault Divorce," p. 1.

65 *"It's so intuitively obvious," says David Blankenhorn* Barbara Vobejeda, "Critics, Seeking Change, Fault 'No-Fault' Divorce Laws for High Rates," *Washington Post,* 7 March 1996, p. A3.

66 *Make marriage harder if you want to* Patchett, in *Women on Divorce,* eds. Kaganoff and Spano, p. 11.

67 *divorce is twice as likely* Winner, *Divorced from Justice,* p. 281. See third Note to page xii for further corroboration.

78 *This is seldom the wife, thanks to the wage gap* Women's wages have remained virtually unchanged, at between one-half and two-thirds of men's wages since the seventeenth century, Phillips, *Untying the Knot,* p. 249. In his history of divorce in the Western world, Roderick Phillips cites the case, reported just before the first World War, of an English woman who had saved twenty years for her divorce. Phillips, *Untying the Knot,* p. 237.

87 *Most child support guidelines are based in large part* Guidelines vary state to state. The most well-known guideline of this type is the Wisconsin Percentage of Income Standard. Based on gross income, the percentages are: one child, 17 percent; two children, 25 percent; three children, 29 percent; four children, 31 percent; five or more children, 34 percent. Minnesota's model, also prominent, is based on net income. It is: one child, 25 percent; two children, 30 percent; three children, 35 percent; four children, 39 percent; five children, 43 percent; six children, 47 percent; seven children, 50 percent. Source: Margaret Campbell Haynes, ed., *Child Support Guidelines: The Next Generation,* U.S. Department of Health and Human Services, Administration for Children and Families, Office of Child Support Enforcement, January, 1994, p. 5.

89 *Custody is still overwhelmingly, though not automatically, awarded to mothers* A 1990 survey of nineteen states conducted by the National Center for Health Statistics found that custody went to mothers in 71 percent of the cases, to fathers in 8.5 percent of the cases, and was shared in 15.5 percent of the cases. (Friends or other relatives received custody in the remainder of the cases.) Jan Hoffman, "Divorced Fathers Make Gains in Battles to Increase Rights," *New York Times,* 26 April 1996, p. B5.

    In almost nine out of ten cases, when parents live apart it is mothers who have custody. Haynes, ed., *Child Support Guidelines,* p. 95.

89 *the more regularly dads see their children* Hoffman, "Divorced Fathers Make Gains in Battles to Increase Rights," p. A1, citing a 1990 Census Bureau report.

90   *a tactic lawyers are not hesitant*   This has been documented in gender-bias reports issues by state courts. Winner, *Divorced from Justice*, p. 49.

90   *the best evidence is that when fathers challenge mothers for custody*   Winner, *Divorced from Justice*, p. 45.

90   *Many states passed joint custody laws and gender-neutral standards*   Winner, *Divorced from Justice*, p. 44.

91   *"Her prosecution of O. J. Simpson"*   "Page Six," *New York Post*, 3 February 1995.

91   *"Men who win custody do not settle into new roles"*   Winner, *Divorced from Justice*, p. 49.

91   *Nor do husbands of working women compensate*   Winner, *Divorced from Justice*, p. 46.

93   *without the legal right to share in decision-making*   Wallerstein and Kelly, *Surviving the Breakup*, p. 310.

Chapter 4 • Money and Work

98   *no researcher was able to replicate Weitzman's findings*   For several studies critical of Wallerstein's findings, see Ahrons, *The Good Divorce*, p. 275.

98   *Dr. Weitzman herself conceded*   Felicia R. Lee, "Influential Study on Divorce's Impact Is Said to Be Flawed," *New York Times*, 9 May 1996.

98   *Current research cites a 20 to 30 percent drop in women's standards of living*   The two-decade "5,000 Families" study conducted by Saul Hoffman of the University of Delaware and Greg Duncan of the University of Michigan showed a 30 percent decline in women's living standards in the first year after divorce, and a 10 to 15 percent improvement for men (Faludi, *Backlash*, p. 88). Those numbers were confirmed by a 1991 Census Bureau study (Faludi, *Backlash*, p. 91). Sociologist Judith Seltzer at the University of Wisconsin, puts the immediate per capita decline in income of households with kids at 21 percent. The latest study, conducted by Dr. Richard R. Peterson of the Social Science Research Council in New York, is that women experience an average 27 percent decline in their standard of living and men a 10 percent improvement. Leland, "Tightening the Knot," p. 73.

98   *only 15 percent of divorcing women are awarded alimony or maintenance*   Even at the turn of the century, when far fewer women worked, only 13 percent of American women requested alimony. In 1916 the percentage of women awarded alimony was only 15 percent. Phillips, *Untying the Knot*, p. 233. That figure has remained relatively constant. Faludi, *Backlash*, p. 95.

98   *Thirty-eight percent of divorced or separated women with kids live in poverty*   Leland, "Tightening the Knot," p. 73.

98   *The demographically "average" American woman*   Naomi Wolf, *Fire with Fire: The New Female Power and How to Use It* (New York: Fawcett Columbine, 1994), p. xxi.

98 *child support is often inadequate or nonexistent* Only about 60 percent of all single mothers have orders of child support, and only about half of these orders are fully paid. Douglas J. Besharov, "How to Help Welfare Mothers," *New York Times* Op-Ed page, 13 November 1996. Nationally, the average court-awarded child support payment is $34 a week. Pollitt, *Reasonable Creatures*, p. 6, confirms that only half of ex-husbands pay it in full.

99 *Weitzman herself observed that even older women who felt economically deprived* Faludi, *Backlash*, p. 1263.

99 *"more an invention of privileged people"* Helen Fisher, *Anatomy of Love: The Mysteries of Mating, Marriage, and Why We Stray* (New York: Fawcett Columbine, 1994), p. 299.

100 *Women looking for jobs in male-dominated fields were considered traitors* Chafe, *Paradox of Change*, p. 109.

100 *A 1936 Gallup Poll asked whether wives should work* Chafe, *Paradox of Change*, p. 116.

101 *In the age group most likely to divorce* Rosin, "Separation Anxiety," p. 16.

101 *today women constitute 46 percent of the work force.* "Few Women at the Top," AP wire, *Washington Post*, 20 October 1996.

101 *Yet women make only 72 cents* Hochschild, *The Second Shift*, p. 249; Pollitt, *Reasonable Creatures*, p. 6; Judith Lorber, *Paradoxes of Gender* (New Haven: Yale University Press, 1994), p. 196. Globally, it's grimmer: a frequently cited United Nations report claims that women do two-thirds of the world's work, receive 10 percent of the income, and own 1 percent of the property. Lorber, *Paradoxes of Gender*, p. 288.

101 *The wage gap figure is based on full-time workers* Susan Faludi, "Statistically Challenged," *The Nation*, 15 April 1996, p. 10. She refers to a Census Bureau statistic that women make 76 percent of what men make.

101 *"If men have defaulted on the pact"* Barbara Ehrenreich, *The Hearts of Men: American Dreams and the Flight from Commitment* (New York: Anchor Books, 1984), p. 173.

101 *Traditional marriage serves its intended economic function* Terry Arendell, *Mothers and Divorce: Legal, Economic, and Social Dilemmas* (Berkeley: University of California Press, 1986), p. x.

102 *Catalyst, a New York–based research company* Kirsten Downey Grimsley, "From the Top: The Women's View," *Washington Post*, 28 February 1996.

102 *"whimsy, unpredictability, and patterns of thinking"* Brownmiller, *Femininity*, p. 16.

102 *Linda Bloodworth-Thomason declares* Judith Lewis, "Comedy Isn't Pretty," *Elle*, February 1996, p. 134.

103 *A USA Today front-page story* Sharon L. Peters, "Some Women on Fast Track Feel Derailed," *USA Today*, 1 February 1996, p. 1.

104 *Despite the handicaps they face in the workplace* Despite feeling that they do not get the pay, benefits, or recognition they deserve, four out

of five of the 250,000 women surveyed by the Women's Bureau of the U.S Department of Labor said that they liked or loved their jobs. Tamar Lewin, "Working Women Say Bias Persists," *New York Times*, 15 October 1994.

104   *Women who stick to their professional guns* Wives employed full-time outside the home do 70 percent of the housework; full-time house-wives do 83 percent. Rix, ed., *The American Woman 1990–91*, quoted in *WAC Stats*, p. 62.

104   *"an indirect way in which the woman pays"* Hochschild, *The Second Shift*, p. 220.

105   *A study by Sarah Fenstermaker Berk* Sarah Fenstermaker Berk, *The Gender Factory: The Apportionment of Work in American House-holds* (New York: Plenum, 1985), pp. 195–96, quoted in Lorber, *Paradoxes of Gender*, p. 191.

105   *"the wife who keeps her household running"* Lorber, *Paradoxes of Gender*, p. 191.

105   *Women's assumption of these duties* Lorber, *Paradoxes of Gender*, p. 189.

105   *"This study shows that women challenge the zero-sum model"* Tamar Lewin, "Women Are Becoming Equal Providers," *New York Times*, 11 May 1995.

105   *No wonder working women with young children* Mary Frances Berry, *The Politics of Parenthood* (New York: Penguin Books, 1993), p. 20.

106   *And while the amount of time fathers spent* Jamieson, *Beyond the Double Bind*, p. 62.

108   *"The low-wage mother . . . is assumed"* Jamieson, *Beyond the Double Bind*, p. 63.

108   *women in upper management are scarce as hen's teeth* As of July 1990, fewer than one-half of 1 percent of top-echelon corporate managers are female. Barbara Ehrenreich, *The Snarling Citizen* (New York: HarperPerennial, 1995), p. 89.

108   *They forfeit good pay* Caryl Rivers and Rosalind C. Barnett, "Are Dual-Income Families Walking Time Bombs?" *Chicago Tribune*, 9 July 1996, p. 11.

108   *Single mother Annie Lamott wrote* Operating Instructions   Annie Lamott, *New York Times*, 1 December 1994.

110   *"No other technique for the conduct of life"* Sigmund Freud, *Civilization and Its Discontents* (New York: Dover Books, 1994), p. 116.

110   *This is not just true of men, or even of Americans* "Women in Work Force," Reuters, *International Herald Tribune*, 2 February 1996, p. 8. Survey of 6,781 women and men conducted by the European Women's Lobby.

111   *The major source of stress for working women* Carol Tavris, "Goodbye, Ozzie and Harriet," review of *She Works, He Works* by Rosalind Barnett and Caryl Rivers, *New York Times Book Review*, 22 September 1996, p. 27.

111 *dual-income families are more stable financially and maritally* Pepper Schwartz, "Me Stressed? No, Blessed," *New York Times*, 19 November 1994.

111 *"it is up to the man to make the* couple's *mark in the world"* Among couples who live together, the more each individual earns, the more satisfied he or she is with the entire relationship; a partner's income does not matter. Once married, however, the wife's income (whether high or low) diminishes in importance. Philip Blumstein and Pepper Schwartz, *American Couples* (New York: William Morrow, 1983), p. 71.

111 *"This makes us think it is not women's working and achieving that causes problems"* Blumstein and Schwartz, *American Couples*, p. 165.

111 *"if money is the key organizing principle"* Hochschild, *The Second Shift*, p. 220.

112 *Blumstein and Schwartz bluntly concluded that except among lesbian couples* Blumstein and Schwartz, *American Couples*, p. 53.

114 *Terry McMillan, the author of* Waiting to Exhale, *gets to the heart of it* Malcolm Gladwell, "Exhale's Fresh Air," *Washington Post*, 23 December 1995.

115 *"it is not always clear whether employed women divorced"* Phillips, *Untying the Knot*, p. 229.

115 *"in many ways, the fact that wives work both benefits and stabilizes marriage"* Hochschild, *The Second Shift*, p. 289.

115 *Dr. William Julius Wilson* David Remnick, "Dr. Wilson's Neighborhood," *The New Yorker*, 29 April and 6 May double issue, 1996, p. 96.

116 *Work reduces the likelihood of divorce* Faludi, *Backlash*, p. 1261.

116 *The happier employed women are with their jobs* Blumstein and Schwartz, *American Couples*, p. 155.

120 *"By simply asserting women's right to enter the labor market"* Ehrenreich, *The Hearts of Men*, p. 151,

121 *"Women lack a positive emotional vocabulary about money"* Wolf, *Fire with Fire*, p. 40.

125 *More than half of the 1,502 working women* Lewin, "Women Are Becoming Equal Providers," *New York Times*, 11 May 1995.

125 *Three out of five working women surveyed in Europe* "Women in Work Force," Reuters, *International Herald Tribune*, 2 February 1996, p. 8.

129 *A 1991 study in* Working Woman *magazine confirmed* *Working Woman*, April 1991. Five hundred recently divorced managers and professionals were surveyed. Ferris State University is in Big Rapids, Michigan.

130 *"Much to my surprise, many of these women said they felt more satisfied"* Ahrons, *The Good Divorce*, 15.

130 *"Powerful working women will almost certainly sustain"* Fisher, *Anatomy of Love*, p. 309.

Chapter 5 • Spirit and Sense of Self

132   *When* New Woman *polled four thousand married readers*   Stephanie von Hirschberg, "The State of Marriage," *New Woman*, June 1995, p. 83.

135   *"the illusory clarity"*   Bateson, *Composing a Life*, p. 199.

136   *They are what historian Carolyn Heilbrun dubs "ambiguous women"*   Heilbrun, quoting Deborah Cameron: "Men can only be men if women are unambiguously women." Carolyn Heilbrun, *Writing a Woman's Life* (New York: Ballantine Books, 1989), p. 21.

137   *"Roles, after all, are not fit aspirations for adults"*   Ehrenreich, *The Hearts of Men*, p. 170.

138   *"like two people at once"*   Jane Shapiro, "This Is What You Need for a Happy Life," in *Women on Divorce*, eds. Kaganoff and Spano, p. 41.

139   *Psychologist Carol Gilligan believes*   Carol Gilligan, *In a Different Voice: Psychological Theory and Women's Development* (Cambridge: Harvard University Press, 1993), p. 42.

142   *"The emotional ground rules immediately shift to self-interest"*   Abigail Trafford, *Crazy Time: Surviving Divorce and Building a New Life* (New York: HarperPerennial, 1992), p. 117.

143   *"Above all other prohibitions, what has been forbidden to women is anger"*   Heilbrun, *Writing a Woman's Life*, p. 13.

143   *"Intended or not, divorce punishes divorcing men and women"*   Riley, *Divorce*, 187.

144   *In an essay on her divorce, writer Ann Patchett harrowingly describes the despair*   Patchett, in *Women on Divorce*, eds. Kaganoff and Spano, p. 9.

145   *In her study on the long-term effects of divorce*   Forty-two percent of the women in Wallerstein's study suffered from chronic depression for more than five years prior to the marriage ending. Three-quarters of the initiators were women. Trafford, *Crazy Time*, p. 141.

145   See *On Death and Dying* by Elisabeth Kübler-Ross (New York: Macmillan, 1991).

147   *"usually women who had been seriously depressed"*   Trafford, *Crazy Time*, p. 141.

147   *Trafford quotes counselor Sharon Baker*   Trafford, *Crazy Time*, p. 266.

147   *Divorce is consistently ranked second*   Arendell, *Mothers and Divorce*, p. 3.

148   *Researchers find that three-quarters of divorcing spouses experience some lingering attachment*   Trafford, *Crazy Time*, p. 130.

151   *"almost all clinical growth takes place in the imperative of unhappiness"*   Augustus Napier, clinical psychologist and director of Atlanta's Family Workshop, quoted in Trafford, *Crazy Time*, p. 11.

161   *In the Children of Divorce project, Judith Wallerstein confirms*   Wallerstein and Kelly write: "The divorce served a useful purpose for many adults, but particularly for the women. More than half of the women arrived at better solutions for living their lives, and half this

group had undergone striking and significant positive changes that appeared to have lifelong implications. Fewer men seemed to have utilized the divorce experience to bring about positive changes in their lives, perhaps because fewer men actually sought the divorce initially." Wallerstein and Kelly, *Surviving the Breakup*, p. 193.

161 *Despite the hurdles posed by psychology* Abigail Trafford quotes researcher Patricia Diedrick's doctoral study at the University of Georgia: "Women face more stressors due to divorce than do men because of gender differences in income, social activity and single parenthood. This had led [some researchers] to conclude that women suffer more as a result of divorce. Yet there is overwhelming evidence that females actually fare better in terms of divorce adjustment than do males. . . . For females a sense of growth in self-esteem appears to result from the divorce. The effects of such changes appear long-lasting." Trafford, *Crazy Time*, p. 163.

161 *"exulting in personal victory is a harshly unfeminine response"* Brownmiller, *Femininity*, p. 212.

164 *"the more I was treated as a woman"* Jan Morris, *Conundrum* (New York: Signet, 1975), p. 166.

166 *By the five-year mark* Faludi, *Backlash*, p. 100.

Chapter 6 • Kids and Moms

169 *Their experiences reflect what studies* Rivers and Barnett, "Are Dual-Income Families Walking Time Bombs?" p. 11. Rivers and Barnett are the authors of *She Works, He Works: How Two-Income Families Are Happier, Healthier, and Better Off* (San Francisco: HarperCollins, 1996). See also Note to page 178.

Summarizing seven national studies of children of divorced parents, sponsored by the National Institute of Mental Health, *USA Today* stated that divorce does not harm children, and that kids who live with divorced mothers actually do better on achievement tests and have fewer problems than those in two-parent homes. *USA Today*, 20 December 1984, cited in Arendell, *Mothers and Divorce*, p. 86.

171 *"We do not have a family-friendly society"* Arlie Hochschild, quoted in Gleick, "Should This Marriage Be Saved?"

172 *single motherhood means financial hardship* Children in households headed by single women account for more than half of all poor children in the United States. Haynes, ed., *Child Support Guidelines*, p. 95. The median 1993 income for households headed by single women was $13,459. Costello and Krimgold, eds., *The American Woman*, p. 403.

A study of the impact of the 1989–90 state child support guidelines conducted by the Women's Legal Defense Fund found that they were reasonably fair to families in which both parents both worked, the custodial parent earned at least two thirds as much as the noncustodial parent, and there were no more than two children. However, the study went on to note, few American families fit that description. Haynes, ed., *Child Support Guidelines*, p. 95.

172 *More than 40 percent of divorced women are not awarded any child support* Trafford, *Crazy Time*, p. 161.

Child support payments amount to only about 10 percent of the income of separated and single mothers, and 13 percent of the income of divorced mothers. Beth Joselow, "Divorced from Reality," *Washington Post*, 22 October 1995.

172 *children are often on their best behavior with the noncustodial parent* Wallerstein and Kelly say this behavior is particularly typical of older children. They write, "The greater their indignation and disapproval [of the divorce and related stresses], the higher the likelihood of angry outbursts directed at both parents, but especially at the parent who sought the divorce. Wallerstein and Kelly, *Surviving the Breakup*, p. 54.

172 *good behavior is often motivated by fears of abandonment* Wallerstein and Kelly, *Surviving the Breakup*, p. 126.

173 *The backlash against these mothers is summed up* Jonathan Alter, "Get Married, Madonna," *Newsweek*, 29 April 1996, p. 51.

173 *"Research shows again and again that poverty and unemployment"* Judith Stacey, "The Father Fixation," *Utne Reader*, September–October 1995, p. 73.

173 *The most recent and comprehensive study* Susan Chira, "Study Says Babies in Child Care Keep Secure Bonds to Mothers," *New York Times*, 21 April 1996, p. A1.

174 *Fifty years of studies show* Rivers and Barnett, "Are Dual-Income Families Walking Time Bombs?" p. 11.

174 *"The issue of child care is really an issue of power"* Berry, *The Politics of Parenthood*, p. 41.

174 *The proportion of "traditional families" in the United States declined* Elizabeth Kolbert, "Whose Family Values Are They, Anyway?" *New York Times*, 6 August 1995, section 4, p. 1.

In 1993 nearly one out of three Hispanic children (32 percent), one out of five white children (21 percent) and nearly three in five black children (57 percent) lived with a single parent, usually Mom. Costello and Krimgold, eds., *The American Woman*, p. 362. Source: Census Bureau, *Marital Status and Living Arrangements, March 1993*, 1994.

174 *By the year 2000, nearly half the population* Ahrons, *The Good Divorce*, p. 212.

The number of children affected by divorce grows by about one million a year. Joselow, "Divorced from Reality."

174 *Sixty percent of American families are headed by a single parent* Kai Erikson, review of *Dubious Conceptions* by Kristin Luker, *New York Times Book Review*, 1 September 1996, p. 12.

174 *"The top priority this nation faces"* Kolbert, "Whose Family Values Are They, Anyway?"

175 *Not until the mid-nineteenth century did courts* Berry, *The Politics of Parenthood*, p. 58.

175 *Mothers who wanted day care had a very hard time* Berry, *The Politics of Parenthood*, pp. 121–22. By 1971 less than 8 percent of white

working women placed their children in day care centers; almost twice as many black working women did so.

175 *That stigma lingers to this day* The United States has no national day care system and no public nursery schools. And, unlike virtually every other Western industrialized nation, it also lacks a national maternity leave policy. Pollitt, *Reasonable Creatures*, p. 8.

175 *these mothers . . . now represent only 3 percent of the population.* Carol Tavris, *She Works/He Works* by Rosalind C. Barnett and Caryl Rivers and *In the Name of the Family* by Judith Stacey, *New York Times Book Review*, 28 September 1996, p. 27.

175 *Full-time homemaking was never an option for the majority* Three-quarters of school-age children now have mothers who work, including 52 percent whose children are under the age of two, and more than one-half of these mothers work full-time. Only one woman in five is a full-time homemaker. Berry, *The Politics of Parenthood*, p. 8.

175 *And a collective amnesia prevails about the dark side of those postwar years* This is discussed at length in Stephanie Coontz, *The Way We Never Were: American Families and the Nostalgia Trap* (New York: Basic Books, 1993).

175 *growing up with bio-mom and bio-dad is no guarantee of psychological well-being* Social scientist Judith Stacey writes: "The claim that intact two-parent families are inherently superior rests exclusively on the misuse of statistics and on the most elementary social science sins—portraying correlations as through they were causes, ignoring mediating factors, and treating small, overlapping differences as gross and absolute." Stacey, "The Father Fixation," p. 72.

175 *It sanctions job discrimination against parents who work* Tamar Lewin, "Men Whose Wives Work Earn Less, Studies Show," *New York Times*, 12 October 1994.

176 *gay people are good parents too, as many studies evidence* "Nearly three decades of research finds gay and lesbian parents to be at least as successful as heterosexuals. Dozens of studies conclude that children reared by lesbian or gay parents have no greater gender or social difficulties than other children, except for the problems caused by homophobia and discrimination." Stacey, "The Father Fixation," p. 73.

176 *"At some deep, queasy Freudian level"* Ehrenreich, *The Snarling Citizen*, p. 46.

177 *Much hand-wringing followed the publication* As is the case with most studies of the children of divorced parents, the subjects were drawn from a group of troubled families who had already sought psychiatric counseling, and the authors never tested their theories on a control group of intact families. More strikingly, the study largely passes by the 45 percent of children who are doing just fine ten years later. More than fine: In Wallerstein's words, "They emerged as competent, compassionate and courageous people." Ahrons, *The Good Divorce*, p. 15.

177 *"People who get divorced probably should get divorced"* Quoted in Rosin, "Separation Anxiety," p. 16.

177   *If children of divorce suffer psychological problems*   Constance Ahrons summarizes the risks of generalizing from Wallerstein and Blakeslees's findings and quotes Joan B. Kelly's review of the current research ("Current Research on Children's Postdivorce Adjustment: No Simple Answers," *Family and Conciliation Courts Review* 31, no. 1 [1993], p. 45): "Overall, the evidence suggests that when children begin the divorce experience in good psychological shape, with close or loving relationships with both parents, their adjustment will be maintained by continuing their relationships with both parents on a meaningful basis." Ahrons, *The Good Divorce*, pp. 275–77

177   *Research seldom compares the effects of divorce*   Arendell, *Mothers and Divorce*, p. 88 footnote.

178   *"The researchers said that if children emerged from the trauma of divorce"*   Barnett and Rivers, *She Works, He Works*, p. 201. The authors cite "Children Are Not Crippled by Divorce Says Australian Study," *Agence France Presse* (May 17, 1993) and Andrew J. Cherlin et al., "Longitudinal Studies on Effects of Divorce on Children in Great Britain and the United States," *Science* 252, no. 5011 (June 7, 1991), pp. 1386–89, in which Cherlin reviews a British study of over 17,000 families.

  Another study was conducted by James L. Peterson and Nicholas Zill, "Marital Disruption, Parent-Child Relationships, and Behavior Problems of Children," *Journal of Marriage and the Family* 48 (1986), pp. 295–307. They find that children in conflict-ridden intact homes are doing worse psychologically than children in single-parent homes.

  A third analysis of the research is Paul Amato and Brian Keith, "Parental Divorce and the Well-being of Children: A Meta-analysis," *Psychological Bulletin* 110, no. 1 (1991), pp. 26–46, which concludes that overall divorce has only a slight negative effect on children.

  In correspondence with the author, Professor Cherlin of Johns Hopkins University emphasizes the presence of conflict, noting that many marriages that end in divorce are not characterized by severe conflict, and that "in this situation, no one has any evidence to show that the children are better off if the parents split." Andrew Cherlin, e-mail correspondence, 16 September 1996.

181   *"If you do these two things"*   Quoted in Trafford, *Crazy Time*, p. 165.

182   *"These two factors differentiate between the children who are and are not damaged by divorce"*   She notes that these two factors have shown up over and over in many different studies, in different populations and in different countries. Ahrons, *The Good Divorce*, p. 127.

183   *"A good divorce does not require that parents share child-care responsibilities equally*   Ahrons, *The Good Divorce*, p. 163.

187   *"One tightly-held secret"*   Ahrons, *The Good Divorce*, p. 9.

188   *Dina's arrangement exemplifies the situation*   Lorber, *Paradoxes of Gender*, p. 157.

189   *A study of eighty divorced mothers*   Arendell, *Mothers and Divorce*, p. 107.

192 *a father's access to his children can be restricted* Arendell, *Mothers and Divorce*, p. 116.

193 *This emotional maturity* Wallerstein and Kelly note that among adolescents in particular, "the emotional and intellectual growth that was catalyzed by the family crisis was impressive and sometimes moving." Wallerstein and Kelly, *Surviving the Breakup*, p. 89.

199 *Statistics bear out the logical assumption* Ahrons, *The Good Divorce*, p. 153, Arendell, *Mothers and Divorce*, p. 107. Almost 80 percent of fathers with visitation rights pay support on time and in full, as do 90 percent of those with joint custody, according to Stuart A. Miller, senior legislative analyst for the American Fathers Coalition, who also points out that about 47 percent of mothers ordered to pay support do not come through. Stuart A. Miller, "The Myth of Deadbeat Dads," *Wall Street Journal*, 3 February 1995, p. A14. Miller cites a 1991 Census Bureau study.

200 *a good relationship with a stepfather can fill the need* Wallerstein and Kelly, *Surviving the Breakup*, p. 200.

200 *The California Children of Divorce project found that good father-child relations* Wallerstein and Kelly, *Surviving the Breakup*, p. 218.

200 *Where disruption occurred in what had previously been a loving relationship* Wallerstein and Kelly, *Surviving the Breakup*, p. 248.

203 *Nationally, 25 percent of divorced fathers make no child support payments* Uncollected child support affects the lives of approximately twenty-three million children, "who, the authors point out, are far less likely to obtain medical care, far more likely to go unsupervised while their mothers work, far less able to afford after-school activities, and far more likely to go hungry at times." Ellen Ruppel Shell, review of *Deadbeat Dads* by Marcia Mobilia Boumil and Joel Friedman, *Washington Post Book World*, 7 April 1996.

203 *Divorced men are more likely to meet their car payments* Faludi, *Backlash*, p. 97.

204 *According to the Office of Child Support Enforcement* $14.1 billion in current support and $30.8 billion in prior years' support was due in 1994. Almost $7.6 billion of the 1994 debt was collected, or 54 percent, but only $2.1 billion, or 7 percent, was collected from the amount owed in years past. Combined, 22 percent of the total amount due was collected, slightly less than the percentage collected in 1993. *Child Support Enforcement Nineteenth Annual Report to Congress for the Period Ending September 30, 1994*, U.S. Department of Health and Human Services, Administration for Children and Families, Office of Child Support Enforcement, p. 43.

204 *Collecting child support is more complicated than it seems* Judges have the authority to enforce awards but are typically reluctant to send deadbeat fathers to jail, the legal consequence of noncompliance. The self-employed are exempt from having their paychecks attached, and administrative obstacles make it almost impossible to collect when the noncompliant parent lives out of state. Winner, *Divorced from Justice*, p. 153.

205   *Three out of four divorced people remarry*   Haynes, ed., *Child Support Guidelines*, p. 13.

206   *"A truly feminist, pro-child divorce reform"*   Rosin, "Separation Anxiety," p. 16.

206   *Rosin notes a distasteful reason*   Rosin, "Separation Anxiety," p. 17.

206   *Forty percent of American children whose parents are divorced*   Harper's Index, *Harper's*, April 1996, p. 8.

206   *The majority of divorced dads leave the parenting to their ex-wives*   In the United States, female-headed families with no spouse present have increased from 11.4 percent of the population in 1970 to 18.1 percent in 1993 (and jumps to 47 percent among the black population); over the same period, the number of male-headed households grew only slightly, from 2.4 percent to 4.3 percent. Summary of "Families in Focus" report co-authored by Judith Bruce, Cynthia B. Lloyd and Ann Leonard, issued by the Population Council in New York City on May 30, 1995, p. 7. The Population Council is an international nonprofit group based in New York that studies reproductive health.

207   *A decade of research shows that remarriage may or may not benefit children emotionally*   Trafford, *Crazy Time*, p. 249.

209   *The incest rate among girls who live with stepfathers*   David Blankenhorn, *Fatherless America: Confronting Our Most Urgent Social Problem* (New York: Basic Books, 1995), p. 40.

Chapter 7 • Sex and Body Image

221   *"Oh, please, Lord, break the curse on women's hearts"*   *Life*, July, 1992.

221   *"The ability of women to participate equally,"*   Jamieson, *Beyond the Double Bind*, p. 61.

221   *not that the Pill is all that safe*   Even at current low and moderate dosages, the risks of taking oral contraceptives include blood clots, liver tumors, and high blood pressure. Some studies have also linked use of the pill to a higher incidence of cancer of the liver, breast, and cervix. Oral contraceptive in-package literature provided by the Ortho Pharmaceutical Corporation, Raritan, New Jersey.

228   *"many of their somatic symptoms"*   Wallerstein and Kelly, *Surviving the Breakup*, p. 306.

233   *"For a woman, getting divorced often heralds a personal sexual revolution"*   Trafford, *Crazy Time*, p. 191.

233   *"power of transgression"*   Heyn, *The Erotic Silence of the American Wife*, p. 37.

233   *They reject the tired premise that only "naughty girls" take the sexual initiative*   Talking about difficulties of letting the lead character in Cybill Shepherd's new situation comedy *Cybill* and her sidekick have sex lives without becoming unacceptable to the viewing public, someone connected to the show—who unsurprisingly insisted on anonymity—said bluntly, "You cannot have a lovable, sexually aggressive woman." Betsy Sharkey, "Good Girls, Bad Girls and Scrambled Signals," *New York Times*, 18 September 1995.

238 *"An ugly woman has no credibility"* Schaef, *Women's Reality*, p. 74.

238 *One's lover's tastes do matter* Brett Silverstein and Deborah Perlick argue that changes in women's roles bring changes in women's bodies: the higher the education level (as in the eras of flappers and Twiggys), the thinner the physical ideal and the more gender-neutral the fashions. Brett Silverstein and Deborah Perlick, *The Cost of Competence* (New York: Oxford University Press, 1995), p. 21.

Chapter 8 • New Relationships

242 *If women were innately ill-equipped to be self-reliant* Heyn, *The Erotic Silence of the American Wife*, p. 141.

242 *married women wouldn't be the most depressed segment of the population* See Note to page 3.

243 *"If I don't have a boyfriend"* Leora Tanenbaum, "Catfights," *New York Press*, 9–15 March 1994, p. 1.

243 *procrustean bed* The dictionary definition is "an arbitrary standard to which exact conformity is forced," after Procrustes, a giant in Greek legend who stretched or shortened captives to fit one of his iron beds. *American Heritage Dictionary*, 1973.

245 *The prospect of solitude no longer terrorizes them* Blumstein and Schwartz's study of thousands of American couples found that women want more time to themselves than men do. Blumstein and Schwartz, *American Couples*, p. 183.

246 *much was made of a 1986 Harvard-Yale study* See first Note to page xii. See also Fisher, *Anatomy of Love*, p. 298.

246 *Critic Katha Pollitt points out* However, the implications of the study are disturbing, as Pollitt goes on to point out: "if the study tells us that women are no longer dependent on marriage for sex or a home or a set of nice dishes, it also tells us that men still want wives who will put husbands first, and that marriage as an institution still favors that desire." Pollitt, *Reasonable Creatures*, p. 8.

246 *Statistics show that three out of four divorced Americans do remarry* Haynes, ed., *Child Support Guidelines*, p. 13. Also Trafford, *Crazy Time*, p. 224.

246 *For women, the typical interval between remarriages* Fisher, *Anatomy of Love*, p. 115.

246 *The remarriage rate is dropping* Samuel H. Preston, "Children Will Pay," *New York Times Magazine*, 29 September 1996, p. 97. Preston is director of the Population Studies Center at the University of Pennsylvania.

246 *"Surprisingly, women are becoming the gypsies"* Schaef, *Women's Reality*, p. 70.

246 *The fact is that both age and gender do affect remarriage rates for women* Trafford, *Crazy Time*, p. 224.

246 *Another factor rendering remarriage less likely is income* Social scientists Blumstein and Schwartz note that in 1977 women between the ages of thirty-five and forty-four with postgraduate degrees and personal incomes above $20,000 were four times as likely to divorce as

women with lower achievement. They also had a 20 percent chance of never remarrying, especially if the woman married at a relatively late age. Blumstein and Schwartz, *American Couples*, p. 76.

248 *No wonder Rule number 31*, Ellen Fein and Sherrie Schneider, *The Rules* (New York: Warners, 1995), p. 144.

252 *"not only to give form to male adventures and female lives"* Heilbrun, *Writing a Woman's Life*, p. 87.

254 *Noting that over 90 percent of all Americans do marry* Fisher, *Anatomy of Love*, p. 298.

255 *Depressingly, about a quarter of all second marriages break up* Trafford, *Crazy Time*, p. 225.

255 *Some 60 to 85 percent of remarriages* Approximately half the marriages in America in 1993 were remarriages for one or both partners. George Feifer, *Divorce: An Oral Portrait* (New York: New Press, 1995), p. 108.

255 *the older the bride and groom* Trafford, *Crazy Time*, p. 241.

255 *Success in remarriage, as Abigail Trafford puts it* Trafford, *Crazy Time*, p. 242.

264 *Nearly 3.7 million American households* Barbara Vobejda, "Cohabitation Up 85 Percent Over Decade," *Washington Post*, 5 December 1996.

264 *One pragmatic reason is that unmarried dual-income couples* Andrea Martin, "Why Get Married?" *Utne Reader*, January–February 96, p. 17.

265 *"This should make for a successful financial partnership"* Blumstein and Schwartz, *American Couples*, p. 77.

268 *In a 1995 poll of four thousand readers* von Hirschberg, "The State of Marriage," p. 83. Ninety-one percent of Egalitarians said their marriages were "great" or "good" (a whopping 61 percent said "great"), compared to 65 percent of both Traditionals and In-betweens; the other categories were "okay," or "bad." A balance of power at home seems to be good for a couple's sex life as well: while 33 percent of respondents overall said they have 'fabulous" sex lives, 42 percent of Egalitarians made this claim.

270 *Marriage itself is becoming less relevant* Looking ahead to the twenty-first century, demographer Samuel Preston observes: "Although sexual activity begins at an earlier age, men and women marry later, separate from one another more frequently and, once separated, are less likely to remarry." The dramatic increase in out-of-wedlock births—33 percent in 1994 compared to 5 percent in 1950—reflects this fact. Preston, "Children Will Pay," *The New York Times Magazine*, 29 September 1996, p. 97.

Index

About the Author

ASHTON APPLEWHITE was the first person to have four books on the *New York Times* bestseller list, and was a clue on *Jeopardy!* as the author of the Truly Tasteless Jokes series. (Who is Blanche Knott?) *Cutting Loose* earned her a place on the board of the Council on Contemporary Families, and landed her on the enemies list of Phyllis Schlafley's *Eagle Forum*. Ashton is also the author of *This Chair Rocks: A Manifesto Against Ageism*, has written for publications like *Harper's*, *Playboy*, and the *New York Times*, blogs at "This Chair Rocks," is the voice of *Yo, Is This Ageist?*, and is a leading voice for a movement to mobilize against discrimination on the basis of age.